"Guidance on the mandatory competencies needs to be both thorough and practical and *Mandatory Competencies: APC Essentials* is the handbook to guide you through. With its comprehensive consideration of the essential content of each competency it will provide candidates with structured revision material and a critical reminder of the many varied ways the mandatory competencies are woven into everyday surveying practice."

Lauren Etebar, *MRICS, APC Mentor*

"The author breaks down the competencies in simple terms within the revision guide for the APC Mandatory Competencies. This enables candidates to fully understand what they need to consider for inclusion within their submissions and at assessment. This professionally written book also acts as a quick guide for Chartered Surveyors to refresh their knowledge."

Michelle Parkes, *MRICS, APC Chair & Assessor*

Mandatory Competencies

Mandatory Competencies: APC Essentials is the first book in a new series designed to cover what any RICS APC (and AssocRICS) candidate or qualified surveyor needs to know about the mandatory competencies.

Written by Jennifer Lemen, author of *How to Become a Chartered Surveyor* and co-founder of one of the UK's market leading APC training providers, Property Elite, this guide is packed full of bite sized information covering:

- Ethics, Rules of Conduct and Professionalism
- Client Care
- Communication and Negotiation
- Health and Safety
- Accounting Principles and Procedures
- Business Planning
- Conflict Avoidance, Management and Dispute Resolution Procedures
- Data Management
- Diversity, Inclusion and Teamworking
- Inclusive Environments
- Sustainability
- Senior Professional Assessment Mandatory Competencies
- Submission and Interview Advice

Practical and concise, with bullet point checklists and real-life examples and diagrams, this handy guide tells you everything you need to know about the RICS mandatory competencies when studying for your APC or your university exams. Relevant to candidates on all APC pathways, including Commercial Real Estate, Residential, Valuation, Quantity Surveying and Construction, Planning and Development, Building Surveying, Land and Resources, this book will also be a handy reference for qualified surveyors and property professionals. This book will also help you to better support your candidates if you are an APC or AssocRICS Counsellor or Supervisor.

Jennifer Lemen BSc (Hons) FRICS is Chartered Surveyor and co-founder of Property Elite, a company providing training and support to RICS APC, AssocRICS and FRICS candidates. She is also an experienced property consultant (Co-director of Projekt), RICS Registered Valuer and RICS Accredited Mediator, with 15 years' experience working in the commercial property sector. She also has academic experience as Senior Lecturer at the University of the West of England, Lecturer at the University of Portsmouth and Associate Tutor at the University College of Estate Management. Her RICS assessment experience includes sitting on final APC interview panels, APC appeal panels and being a lead APC preliminary review assessor. She also assesses written AssocRICS submissions.

APC Essentials

APC Essentials is a series of books providing guidance to RICS APC and AssocRICS candidates on various mandatory and technical competencies, relevant to each individual pathway.

Whilst this series is structured to cover the full scope of APC competencies, AssocRICS candidates will still find the content and knowledge helpful for the requirements of their assessment.

Mandatory Competencies
APC Essentials
Jennifer Lemen

For more information about this series, please visit: www.routledge.com

Mandatory Competencies

APC Essentials

Jennifer Lemen
BSc (Hons) FRICS

Routledge
Taylor & Francis Group

LONDON AND NEW YORK

Designed cover image: Bobby Darragh and Abigail Blumzon

First published 2024
by Routledge
4 Park Square, Milton Park, Abingdon, Oxon OX14 4RN

and by Routledge
605 Third Avenue, New York, NY 10158

*Routledge is an imprint of the Taylor & Francis Group, an
informa business*

British Library Cataloguing-in-Publication Data
A catalogue record for this book is available from the
British Library

ISBN: 978-1-032-56778-5 (hbk)
ISBN: 978-1-032-44862-6 (pbk)
ISBN: 978-1-003-43711-6 (ebk)

DOI: 10.1201/9781003437116

Typeset in Times New Roman
by Apex CoVantage, LLC

This book is dedicated to those who have been a part of my journey to becoming a Chartered Surveyor. My family: Mum and Dad, Granny and Grandad Birmingham, Granny and Grandad Scotland. My business partner, friend and mentor through life: Rachel Saint. My wife and soulmate, Frankie, and the Amat family. And not forgetting my two best canine friends, Ted and Ben. I wouldn't be here without you all. Thanks also go to Phil Page for proofreading and editing my final text.

Contents

Figures

Foreword

After the success of *How to Become a Chartered Surveyor*, it became clear to the author that RICS APC and AssocRICS candidates are struggling to understand the mandatory competencies they need to achieve.

These are fundamental business skills; sometimes they are written down in textbooks and industry guidance. Otherwise, the only way to learn is to pick up skills from other surveyors in the office and on-site.

The author was able to do this through completing a five-year part-time course in Property Management & Investment at the University of the West of England. Alongside this, the author was employed as a trainee surveyor at Seagrove & Lambert Surveyors in Bristol on a day release basis. Working in a small firm enabled the author to pick up a range of business skills, covering most of the mandatory competencies discussed in this book.

The author later accepted a role at Rapleys and converted her mode of study to distance learning without attendance. This allowed her to take on more responsibilities and gain wider work experience, again exploring a wide range of core business skills.

The author graduated with a First Class degree and an overall average of 91%. She was also awarded the RICS South West Regional Prize for high achievement.

A year later, the author was proud to qualify as Chartered Surveyor through the Royal Institution of Chartered Surveyors (RICS) Assessment of Professional Competence (APC). This process was not easy, with the intense combination of work and studying before preparing for a final professional interview.

After qualifying as Chartered Surveyor, the author proceeded through the career ranks before leaving to set up in practice with her business partner and career-long mentor, Rachel Saint DipSurv

MRICS. Alongside setting up an RICS regulated surveying firm, Projekt, the pair founded a highly successful training and support company for prospective Chartered Surveyors, Property Elite.

The author also attained a higher level of surveying recognition, becoming a Fellow of the RICS (FRICS) and Accredited Mediator, whilst also lecturing and supervising dissertation students at various universities.

The importance of the mandatory competencies has been emphasised throughout the author's career, including running, managing and leading a business, communicating with others and acting ethically and professionally. The author is also proud to have trained and mentored countless candidates to become ethical, professional and competent AssocRICS, MRICS and FRICS professionals.

With the advent and necessity of home-based working in recent years, many surveyors now miss out on vital work-based mentoring and shadowing. This book hopes to fill a gap and to pass on some of this knowledge to the next generation of Chartered Surveyors. Thanks must go to all of the Chartered Surveyors who have contributed to this book, including Don Lemen FRICS and consultants within the fantastic Property Elite team.

The author hopes that you will be inspired to become a better property or construction professional by the end of this book!

Chapter 1

Introduction

Introduction

APC Essentials is a series of books guiding RICS APC and AssocRICS candidates on various mandatory and technical competencies relevant to each sector-specific pathway.

Whilst this series is structured to cover the full scope of APC competencies, AssocRICS candidates will still find the content and knowledge helpful for their assessment requirements.

The series is split into several individual books, starting with this edition on the mandatory competencies and three senior professional assessment mandatory competencies. Other books will cover the sector-specific pathways and related technical competencies.

Overview of the APC Assessment

All APC candidates, irrespective of their chosen route, must undergo assessment via a written submission and an online interview.

The online interview is based on your written submission and lasts one hour. The final decision as to whether you become MRICS is solely based on your performance in this interview.

All candidates, including those undertaking the senior professional, specialist and academic assessments, must choose one of 22 sector pathways which most closely reflects their experience, role and knowledge:

1. Building Control;
2. Building Surveying;
3. Commercial Real Estate;

DOI: 10.1201/9781003437116-1

4. Corporate Real Estate;
5. Environmental Surveying;
6. Facilities Management;
7. Geomatics;
8. Infrastructure;
9. Land and Resources;
10. Management Consultancy;
11. Minerals and Waste Management;
12. Personal Property/Arts and Antiques;
13. Planning and Development;
14. Project Management;
15. Property Finance and Investment;
16. Quantity Surveying and Construction;
17. Research;
18. Residential;
19. Rural;
20. Taxation Allowances;
21. Valuation;
22. Valuation of Businesses and Intangible Assets.

What are the APC Competencies?

Competencies are the individual mandatory ('soft') and technical ('hard') areas where you must demonstrate your skills, knowledge and experience. The competency requirements for each separate pathway are set out in the relevant pathway guides, for example, Corporate Real Estate, which can be downloaded from the RICS website.

The requirements of each specific competency are divided into three levels. The level of competence (either level 1, 2 or 3) required will depend on your chosen pathway and competencies. For example, all candidates must satisfy the mandatory competency, Ethics, Rules of Conduct and Professionalism to level 3. The pathway guides set out the individual requirements of each level and each competency, e.g. required knowledge for level 1 and required activities or advice at levels 2 and 3, respectively.

This series of books is based on the competency descriptors in each relevant pathway guide. You can use the books to ensure they choose competencies reflecting your knowledge and experience. You can also use the books as a guide to level 1 knowledge and level 2 and 3 examples to write up in the summary of experience and ensure that the case study includes sufficient competency-based knowledge, experience and advice.

You should continually refer to the competency descriptors and these books to ensure that what you write in your submission is accurate, relevant and sufficiently detailed. This is because your submission will guide potential areas of questioning in your final assessment interview.

When you prepare for your final assessment interview, this series of books will provide an essential source of knowledge to supplement revision, mock interviews and other preparation.

What Are the Three Competency Levels?

The three competency levels are defined as follows:

- Level 1 – Knowledge and understanding – This requires candidates to explain their knowledge and learning relevant to the description included in the competency guide. This could be through academic learning, e.g., degree-level courses, CPD or on-the-job training. However, candidates should avoid too much repetition as their CPD record will be set out separately. Generally, level 1 will be met by including a brief list of knowledge or relevant topics, given the limitations of the overall word count. These books will guide candidates on each competency's scope, depth and breadth;
- Level 2 – Application of knowledge and understanding – Demonstrating level 2 relates to a candidate applying their level 1 knowledge and understanding in practice, i.e., through practical work experience. This relates to 'doing', rather than just knowing the theory and fundamentals. Candidates should refer to specific projects, instructions or examples to demonstrate their practical activities and experience, including their role and relevance to the competency description. These books will guide candidates on relevant experience to include within each competency;
- Level 3 – Reasoned advice and depth of knowledge – Expanding on level 2 work experience or tasks to provide reasoned advice to clients. This involves considering the options available and offering recommendations and advice on solutions. These books will guide candidates on relevant advice to include within each competency.

Assessors are trained to begin questioning at the highest level declared, with supporting level 1 knowledge-based questions potentially being asked to explore the justification for your advice or actions. You should ensure they are familiar with any examples in levels 2 and 3, as these should form the basis of your answers.

You should not be asked questions on competencies you have not selected or at levels beyond those stated, i.e., you will not be expected to give reasoned advice (level 3) if you have declared a competency only to level 2 (acting or doing).

All pathways have the same mandatory competencies. However, each pathway will have a different set of technical competencies, split into core (primary skills) and optional (additional skills) technical competencies.

What Are the Mandatory Competencies?

Common to all pathways, the 11 mandatory competencies comprise business skills. This means that they need to be demonstrated by all candidates, irrespective of role, sector or position. They aim to show you can work with others, manage your workload, and act ethically and professionally.

The mandatory competencies include:

- Ethics, Rules of Conduct and Professionalism (level 3);
- Client Care (level 2);
- Communication and Negotiation (level 2);
- Health and Safety (level 2);
- Accounting Principles and Procedures (level 1);
- Business Planning (level 1);
- Conflict Avoidance, Management and Dispute Resolution Procedures (level 1);
- Data Management (level 1);
- Diversity, Inclusion and Teamworking (level 1);
- Inclusive Environments (level 1);
- Sustainability (level 1).

You can also pursue some of the mandatory competencies to higher levels as part of your core or optional competency choices on specific pathways.

This book will cover the 11 mandatory competencies and provide excellent baseline knowledge of what you need to know to ace your APC.

If you are enrolled on the senior professional assessment, there are three additional mandatory competencies that you must achieve:

- Leadership (level 2);
- Managing People (level 2);
- Managing Resources (excluding human resources) (level 2).

What Are the Technical Competencies?

The technical competencies relate to the candidate's specific area of practice, rather than being more general business skills. They are split into core and optional competencies, depending on the specific requirements of each pathway. You should pay particular attention to the requirements of the core competencies, as these must be demonstrated to the required level. This may require you to seek additional work experience or secondment to fill gaps in experience, e.g., Planning and Development pathway candidates may need to identify other Valuation experience to fulfil this as a core technical competency to the required level.

Subsequent books will cover the technical competency requirements of various pathways, focusing on the core technical competencies but with additional supporting guidance on different optional technical competencies.

Structure of This Book

Each chapter of this book will focus on a different mandatory competency. The content will be relevant to all pathways, with competency-specific requirements highlighted.

The structure of each chapter will be as follows:

- A summary of the RICS competency requirements (taken from the RICS pathway guides);
- What candidates need to know at level 1, including legislation, RICS guidance and hot topics;
- Examples of how candidates can achieve level 2 (doing);
- Examples of how candidates can reach level 3 (advising);
- Sample APC interview questions on each level.

Conclusion

Continue reading Chapter 2 to explore the full breadth and depth of the RICS APC mandatory competencies.

Remember that this book is accurate as of the publication date. You must conduct your research to fact-check any updates or new hot topics following this date. There is a range of excellent resources online, but ensure you use those that are accurate, reliable and written by a credible source.

Chapter 2

Ethics, Rules of Conduct and Professionalism

Introduction

Acting professionally and ethically is at the heart of what it means to be a Chartered Surveyor. In this chapter, we look at the requirements of the RICS concerning the Ethics, Rules of Conduct and Professionalism competency.

This is a mandatory RICS APC competency required to level 3. You must write your summary of experience, including a statement at levels 1, 2 and 3. You must consider the ethics of your conduct and reasoned advice to clients and ensure that you act and advise in a robust, professional and moral manner.

The RICS pathway guide sets out the relevant knowledge (level 1), practical application (level 2) and reasoned advice (level 3) for this competency. However, you do not need to evidence everything listed, as this will depend on your experience and role.

This competency is not simply about knowing how to act ethically or being aware of the relevant RICS guidance; it is about the reasoned application of ethics to the decisions and actions you take within your personal and professional life.

You will also need a robust understanding of the legal system of the countries that you work in. If you work across multiple countries, you must know about the legal system and relevant legislation in each. For example, a surveyor working in England and Scotland must know legal systems and legislation.

In the final assessment interview, an incorrect or unethical answer concerning the Ethics, Conduct Rules and Professionalism competency will constitute an automatic failure, irrespective of your performance in the rest of the interview.

DOI: 10.1201/9781003437116-2

APC candidates must demonstrate a strong understanding and application of ethics and professionalism through the RICS online Professionalism Module. AssocRICS candidates do not have to write up this competency in their summary of experience. However, they do have to pass the Professionalism Module before submitting and demonstrate ethical and professional behaviour in the submission as a whole.

Once qualified as AssocRICS or MRICS, surveyors must undertake CPD in relation to this competency every three years to satisfy the mandatory requirements of RICS.

Most surveyors will initially be introduced to ethics and professionalism in a module at university (or another academic institution). During any workplace induction, ongoing training and CPD that a surveyor undertakes, this will be built on. Ethics is an area for continued focus throughout a surveyor's career, not just something that should be considered once at the outset.

This chapter will explain what you should know in relation to Ethics, Rules of Conduct and Professionalism, covering the main requirements of level 1 and how to apply them at levels 2 and 3.

UK Legal System

This section on the UK legal section is reprinted from *Real Estate: The Basics* (Forsyth & Wilcox, 2022).

Why is it so important to understand real estate law? Whilst surveyors do not necessarily need the depth and detail of knowledge required by a lawyer, they still need to understand the legal context in which they and their clients operate. Surveyors must be able to recognise potential legal issues at an early stage; they may need to explain legal concepts to clients, and they may need to brief solicitors. They also need to know when to seek legal advice.

Let us look at some examples in which a real estate practitioner may need to understand and apply the law:

- Liaising with solicitors during a sale and purchase;
- Checking boundaries in title deeds and at the Land Registry if there is a dispute;
- Identifying a neighbour's right of way across a site proposed for purchase;

- Recognising that a covenant may prevent site development. Legal problems can be a deal-breaker in a property transaction, making property management challenging.

Solicitors do not usually visit the properties they are dealing with, so the surveyor or agent may be the 'eyes on the ground', able to flag up potential issues that need further investigation before they become a significant concern – and expense – for the client. Understanding legal basics can therefore help a surveyor to provide an excellent service to their client; it is part of being an expert, a professional, and – significantly – it can help the practitioner to avoid the many legal pitfalls that can trap the uninformed.

A helpful starting point is to consider the law itself. What do we mean by the law, and what do we mean when we talk about something being legal?

There are three types of law that real estate practitioners need to be aware of:

- Common law;
- Equity;
- Legislation.

We will look at each one in turn.

Common Law

In England and Wales, common law is the law that has evolved over centuries based on the decisions of the courts. These decisions may set legal precedents, creating rules and principles that become law. Criminal courts are beyond the scope of this book, although there are crimes concerning real estate, for example, in relation to health and safety and money laundering. For our purposes, it is the civil courts that we are concerned with when studying land law. When we use the terms 'legal' and 'in law', we refer to the common law. Courts are organised in a hierarchy that enables appeals to be made to a higher court if a party does not accept the judge's decision to a dispute. The general rule is that the lower courts, for example, the County Court, must follow the decisions made by the courts further up in the hierarchy – the High Court, the Court of Appeal, and the Supreme Court. Each is required to follow the decisions of the courts above it. Historically, the House of Lords was the highest court, but the creation of the Supreme Court in 2009 heralded a clearer separation of Parliament and the judiciary.

Equity

Equity originated in the Courts of Chancery and has also evolved over centuries. The role of equity is to alleviate the harshness that can characterise the common law. Equity is therefore founded on values of fairness and justice, and it can be recognised by its maxims or principles. Examples of these equitable maxims are: 'equity regards as done, that which ought to be done', and 'he who comes to equity comes with clean hands'. Equity will therefore have regard to the conduct of the parties when considering what is fair and just. We will see some examples of equity in practice when we look at aspects of land law in more detail later on. Equity also offers remedies in some disputes, which may be more appropriate than common law remedies. For example, a court may award damages in common law – financial compensation for a breach of a right – whereas equity might grant an injunction to prevent the breach from occurring. An important distinction is that a claimant who successfully proves the breach in court will be entitled to common law remedies such as damages. This contrasts with equitable remedies such as an injunction (a court order that either prohibits someone from doing something or requires them to do something specific), which is discretionary. This means that the court can decide whether to grant the remedy to the claimant, and the claimant's conduct may influence the court's decision. We have seen that equity is based on principles of fairness. Equity can prevent injustice by offering an equitable interest to someone who intended to create a legal (i.e., a common law) interest but failed to do so because they did not follow the procedure required by common law. Equity also recognises certain rights in land – such as restrictive covenants – that only have limited application at common law. Initially, common law and equity were dealt with by different courts, but these functions were combined into one court system in the 1870s. A court can now decide any issue, regardless of whether it relates to common law or equity. However, although the actual courts have merged, common law and equity remain separate. If the common law and equity conflict, equity prevails, meaning it takes priority over common law.

Legislation

Parliament creates legislation in the form of statutes or Acts of Parliament. These can influence both common law and equity. For example, the Law of Property Act 1925 tells us which rights and interests in land can be legal and, consequently, which rights and interests can only be equitable. It is the role of judges to interpret and apply legislation to the legal disputes that reach the courts. The courts' decisions create case law, and you will see the importance of the courts' decisions in all areas of the law. A helpful tip concerns case law: when discussing a court case, legal professionals do not say 'versus'; they say 'and'. To take an example of a case referred to later, we write 'Buckland v Butterfield', but we say: 'Buckland and Butterfield'. It is usual to refer to the year in which the court made its decision or in which the case appeared in the official law reports and to provide the full case citation in an academic assignment or a professional report. This ensures that the reference is to the correct decision as a dispute may progress through a succession of courts to the Supreme Court over several years, with each court making a different decision.

RICS Regulation

Candidates must understand the importance of the RICS and how it regulates the surveying profession. This is not simply background information; it is knowing the institution that you want to gain membership to and being aware of the framework in which you practice.

The Royal Institution of Chartered Surveyors (RICS) is the global governing body for Chartered Surveyors. The RICS was formed initially as the Surveyors Club in 1792 and developed over time into a professional association representing surveyors and the property profession. This reflected the rise of industrialisation and expansion of infrastructure, housing and transport, which required advice from professionals and created a need for some form of regulation.

The Surveyors' Institution was founded on 15th June 1868 and later incorporated by Royal Charter on 26th August 1881. On 27th October 1930, the name was changed to The Chartered Surveyors' Institution, which became what we now know as the RICS on 3rd July 1947. The motto of the RICS is 'Est modus in rebus' or 'There is measure in all things'.

The RICS Royal Charter sets out the fundamental difference between a surveyor and a Chartered Surveyor. It incorporates the Original Royal Charter (dated 26th August 1881) and Supplemental Royal Charter (which amends the Original Royal Charter from time to time, the latest changes being made in February 2020).

The Royal Charter means that changes to the RICS constitution, known as the Bye-Laws, must be approved in a two-step process. Firstly, they must be approved by a majority members' vote at a general meeting. Secondly, the Privy Council, part of the UK Government, must ratify them.

The Royal Charter states that 'Chartered Members', i.e., Chartered Surveyors, may only be Fellows (FRICS) and Professional Members (MRICS) of the RICS. They also give certain firms the right to use the designation Chartered Surveyors. Chartered Surveyors and firms of Chartered Surveyors are required by the Royal Charter to follow the RICS Bye-Laws and Regulations.

The Royal Charter states that the aim of the RICS is 'to maintain and promote the usefulness of the profession for the public advantage in the United Kingdom and in any other part of the world' (RICS, 2020d). In particular, this focuses on protecting the public (consumers and business clients) and upholding the reputation of the RICS within society. RICS also influences policy, sets standards, accredits professionals and provides quality assurance across the industry.

The Royal Charter sets a gold standard of excellence and integrity for the industry, mainly because Royal Charters remain in demand from various professions. Other UK professionals with a Royal Charter include Chartered Accountants, Chartered Managers, Chartered Psychologists, Chartered Structural Engineers, Chartered Members (MCIOB) of the Chartered Institute of Building (CIOB) and Chartered Town Planners.

The fact that RICS has a Royal Charter means that the Chartered Surveying profession operates under a model of self-regulation. This means that there is no Government regulation of Chartered Surveyors. Instead, they are internally regulated by RICS through the Bye-Laws and RICS Regulation.

The Royal Charter means that the Government should be confident that Chartered Surveyors are regulated appropriately and diligently, in line with the principles of better regulation: Proportionality, Accountability, Consistency, Targeting and Transparency. These are set out by the UK Cabinet Office's Better Regulation Commission and are adopted by RICS in their regulatory model.

The ability to self-regulate is beneficial because it avoids the time and cost of the Government introducing and maintaining appropriate legislation. Effectively, there is no current need to legislate because the Government has historically been confident that through the Royal Charter, RICS has internally regulated at arm's length, as a business and in line with modern working practices. Some parts of the industry are subject to or may become subject to external regulation. An example is the Government proposals for the Regulation of Property Agents (ROPA).

The RICS Bye-Laws (RICS, 2020b) and RICS Regulations (RICS, 2020c) were last updated in February 2020. Whilst the Bye-Laws set out more general regulatory principles and procedures, the Regulations set out specific detail relating to the operation of the RICS as a regulatory body.

Together, these set out the regulatory requirements of the RICS, including issues such as membership eligibility, registration of firms, use of designations, subscriptions and fees, conduct, powers and governance structure. The way that RICS regulates is further guided by written Governance Procedures and Processes (also known as Standing Orders) (RICS, 2020a).

The key takeaway from the Bye-Laws and Regulations is that Chartered Surveyors must conduct themselves in a 'manner befitting membership of RICS'.

RICS Structure

RICS has been subject to various independent reviews over the years.

The first of these, in 2005, the Carsberg Review, reviewed the self-regulating role of RICS. This led to the introduction of the Rules of Conduct, which at the time were split into separate Rules for Members and for Firms. It also segregated the regulatory (Regulatory Board) and representational (Governing Council) roles within RICS.

In 2008, a decision at the RICS AGM led to changes to the Royal Charter and Bye-Laws. This established a new corporate governance model, requiring RICS to be run more like a business.

However, 2021 saw the Levitt Review into treasury management issues at RICS during 2018 and 2019. This was closely followed by the Bichard Review, an independent review commissioned by the RICS

Governing Council into the purpose, governance and strategy of RICS. Thirty-six recommendations were made, including separating commercial activity from other activities at RICS, Matrics being reinvigorated, a diversity and inclusion panel being established, the improved use of technology, the development of RICS' thought leadership role (i.e., creative and original thinking from experts leading their industry or sector), five-yearly independent reviews unto the effectiveness of RICS, a customer service improvement programme and a revised RICS corporate structure.

At the highest level, the RICS is led by the Presidential Team. They work with the CEO and Executive Team to shape the future of RICS. The 2023 President is Ann Gray, supported by the President-Elect and Senior Vice President. Members of the Presidential Team have one-year terms, progressing from Senior Vice President through to President-Elect and then President. They also sit on the Governing Council, with the President as Chair.

The RICS' senior leadership team (Executive Team) comprises a Chief Executive Officer (CEO) (from June 2023, this was Justin Young) and various Executive Directors. They are responsible for the day-to-day running of RICS and implementing the decisions of Governing Council, RICS Board and Standards & Regulation Board.

The structure of RICS, as of October 2023, will be explained in turn below.

Governing Council (GC) is the governing body of RICS under the Royal Charter. The role of GC is to set the strategy and vision of RICS. Underneath GC are the RICS Board and the Standards & Regulation Board.

The RICS Board delivers the RICS business plan and oversees operations, excluding standards and regulation. The Board is supported by the Membership Services Committee, Knowledge & Practice Committee, Audit, Risk Assurance & Finance Committee and Nominations & Remuneration Committee.

The Membership Services Committee sits over the World Regional Boards (WRBs). These are responsible for the strategy and business plan for the five world regions: Americas, Middle East & Africa, Asia Pacific, Europe and UK & Ireland.

The Standards & Regulation Board exercises the regulatory function of RICS, including strategy, governance, structure, policy and operational oversight in the public interest. It is accountable to GC, but not specifically directed by them.

The Board also oversees RICS professional standards, entry and admission to the profession, education and qualification standards, regulatory operations, regulatory schemes (such as the Firm Registration, Valuer Registration, Designated Professional Body Scheme and Client Money Protection Schemes), dispute resolution services and Regulatory Tribunal (plus their associated rules). It is independently led and comprises both non-members and Members of RICS.

Where a concern or complaint is raised about a Member or Regulated Firm, the Head of Regulation will consider disciplinary action. This will only be pursued if the matter is serious, evidenced and in the public interest to consider. Serious cases can be referred to a disciplinary panel, which will conduct a hearing and impose appropriate sanctions, including costs being awarded against the Member.

Less serious cases are dealt with by Regulatory Compliance (Consent) Orders (where a Member admits allegations) or referral to a Single Member of the Regulatory Tribunal to determine the sanction (where a Member does not admit allegations). Disciplinary action (sanctions) can include fixed penalty notices, Regulatory Compliance (Consent) Orders, suspension and expulsion.

All RICS Regulated Firms must complete an annual return, including providing information on the following:

- Business type;
- Staffing;
- Statutory regulated activities, e.g., providing financial services;
- Nature of clients;
- Complaints Handling Procedure (CHP);
- Professional Indemnity Insurance (PII);
- Contact officer.

A firm holding clients' money must register with the Client Money Protection Scheme and pay a separate annual Regulatory Review fee.

If a firm's annual return raises any red flags or substantial risks, RICS may conduct an online Desk-Based Review (DBR). A report with actions and recommendations is sent to the firm, which must respond within 28 days. RICS may conduct a physical Regulatory Review Visit to the firm if serious concerns remain.

The Bichard Review also introduced six professional groups: Construction, Valuation, Commercial Property, Residential Property, Land & Natural Resources and Building Surveying & Building

Control. Each will have a Chair who is also a member of the Knowledge and Practice Committee.

Two other bodies to be aware of are Matrics and LionHeart. Matrics are locally organised bodies that support early career-stage surveyors. They run various low-cost or free events and provide an excellent source of networking and support. RICS LionHeart is an independent charity that supports RICS professionals and their families.

RICS Rules of Conduct

Acting ethically and professionally gives clients and the public confidence in the advice that Chartered Surveyors provide.

This is emphasised by RICS research, which stated that

> high ethical standards should be seen as part of an employer organisation's good governance and through this, its competitive advantage, because offering a high-quality service to the public will raise the profile of the employing organisation (which is likely to be reflected in its balance sheet) and thus enhance the reputation of the entire surveying profession as well as that of its employees.
>
> (Plimmer, et al., 2009)

The RICS sets ethical and professional requirements for Chartered Surveyors in the Rules of Conduct, which both Members and Firms must adhere to.

The RICS Rules of Conduct were updated in February 2022, replacing the former Rules of Conduct for Firms and for Members (separate editions). They apply to Members and Firms globally, not just in the UK.

There are five Rules, supported by a list of example behaviours to help Members and Firms to comply.

The five Rules are summarised as follows:

- Rule 1 – Members and Firms must be honest, act with integrity and comply with their professional obligations, including obligations to RICS. Example behaviours include not being improperly influenced by others and being transparent with clients about fees and services;
- Rule 2 – Members and Firms must maintain their professional competence and ensure that services are provided by competent individuals with the necessary expertise. Example behaviours include only

undertaking work where a Member has the relevant knowledge, skills and resources and undertaking sufficient CPD each year;
* Rule 3 – Members and Firms must provide good-quality and diligent service. Example behaviours include understanding clients' needs and objectives before accepting work and communicating clearly with clients;
* Rule 4 – Members and Firms must treat others respectfully and encourage diversity and inclusion. Example behaviours include treating others courteously and respectfully and developing an inclusive workplace culture;
* Rule 5 – Members and Firms must act in the public interest, take responsibility for their actions and act to prevent harm and maintain public confidence in the profession. Example behaviours include managing professional finances responsibly and responding to complaints promptly, openly and professionally.

Appendix A of the Rules of Conduct lists the core professional obligations of Members and Firms to RICS. For Members, these include cooperating with RICS, complying with CPD requirements and providing information promptly to the Standards & Regulation Board. For firms, these include publishing a written Complaints Handling Procedure, ensuring adequate Professional Indemnity Insurance is held and cooperating with RICS.

Only serious breaches of the Rules are likely to result in disciplinary action. RICS states that minor breaches will be handled through self-correction or revised firm-wide processes.

RICS is also a member of the International Ethics Standards Coalition (IESC), alongside many other professional and governing bodies. The IESC publish the International Ethics Standards (IES), which provide a global set of ethical principles for property-related professions.

The fundamental principles of IES are:

* Accountability;
* Confidentiality;
* Conflict of interest;
* Financial responsibility;
* Integrity;
* Lawfulness;
* Reflection;
* Standard of service;
* Transparency;
* Trust.

RICS Members must report to RICS any unethical or unprofessional behaviour or practices of other Members. This duty sits alongside any reporting or whistleblowing procedures set out by a surveyor's employer or other organisations they work with or for.

When reporting a potential issue to RICS, surveyors need to consider their duty of confidentiality towards any clients, which could include anonymising confidential information or seeking the client's permission to disclose information.

APC and AssocRICS candidates should be aware of their duty of confidentiality to clients and third parties when writing their case studies. RICS provides guidance on this in the respective Candidate Guides. Candidates will need their client's consent to use the case study. If this is not given, the candidate must redact identifiable information and reword the confidentiality statement accordingly. Candidates also need to consider the Data Protection Act 2018 requirements concerning the level of identifiable detail they state in their case study.

Supplementary RICS Guidance

Following the Bichard Review, the RICS has overhauled the different categories of professional guidance. Existing guidance, e.g., Professional Statements (mandatory), Guidance Notes (best practice) and Information Papers (broader information or context), is being replaced or updated with new types of guidance as time goes on, and all new guidance is published in the new forms.

Professional guidance is now split into Professional Standards and Practice Information.

Professional Guidance sets requirements or expectations for RICS Members and Regulated Firms about how they provide services or the outcomes of their actions. Where requirements are mandatory, guidance will use 'must', whereas best practice will use 'should'.

Surveyors must have a robust understanding of relevant guidance. This is because, in regulatory or disciplinary proceedings, RICS will take account of relevant Professional Standards in deciding whether a Member acted appropriately and with reasonable competence.

Practice Information, conversely, supports the practice, knowledge and performance of providing surveying services. Again, surveyors must robustly understand relevant Practice Information to ensure they provide the highest service standards.

Chartered Surveyors must also adhere to their ethical and moral values and those set out in their employer's policy and guidance.

RICS CPD Policy

All RICS members (AssocRICS, MRICS and FRICS) must undertake at least 20 hours of CPD each year. This must be completed by 31st December and recorded online (or using the RICS mobile app) by 31st January. At least ten hours must be formal (structured) CPD, and every three years, CPD relating to ethics must be undertaken.

Formal CPD includes more than just formal training courses; it includes anything with a clear learning objective or outcome. Examples include online training, webinars, structured discussions, technical authorship and undertaking academic courses.

Certain activities will not count as CPD, including networking, social events, committees and clubs, which are not linked to a Member's professional role.

Firms must ensure that staff training is structured and promotes the provision of professional work to the required standards of skill, care and professionalism. This will likely include a structured training programme, appraisal system and support for personal development. Firms must declare that this is in place as part of their Annual Return to RICS.

Conflicts of Interest

To act ethically and professionally, Chartered Surveyors need to ensure that they are not influenced by any perceived or actual conflicts of interest.

The RICS set out mandatory requirements relating to conflicts of interest in the Professional Standard Conflicts of Interest (1st Edition). This relates to the actions of Firms and Members and provides guidance on how to comply with the requirements of the Rules of Conduct.

The RICS states that

the most important reason for avoiding conflicts of interest is to prevent anything getting in the way of your duty to advise and represent each client objectively and independently, without regard to the consequences to another client, any third party, or your interests and that the clients and in turn, the public can be confident you are doing so.

(RICS, 2017)

A conflict of interest is defined as (RICS, 2017):

- Party conflict – a situation in which the duty of an RICS Member (working independently or within a non-Regulated Firm or a Regulated Firm) or a Regulated Firm to act in the interests of a client or other party in a professional assignment conflicts with a duty owed to another client or party in relation to the same or a related professional assignment;
- Own interest conflict – a situation in which the duty of an RICS Member (working independently or within a non-Regulated Firm or a Regulated Firm) or a Regulated Firm to act in the interests of a client in a professional assignment conflicts with the interests of that same RICS Member/Firm (or in the case of a Regulated Firm, the interests of any of the individuals within that Regulated Firm who are involved directly or indirectly in that or any related professional assignment);
- Confidential information conflict – a conflict between the duty of an RICS Member (working independently or within a non-Regulated Firm or a Regulated Firm) to provide material information to one client, and the duty of that RICS Member (working independently or within a non-Regulated Firm) or of a Regulated Firm to another client to keep that same information confidential.

Furthermore, where a surveyor has a personal (own interest) conflict of interest and is acting in an agency capacity, they must declare the matter in writing at all stages of the process in line with Section 21 of the Estate Agents Act 1979. This includes reference to the personal interest in the marketing particulars and memorandum of sale. Personal interests could include where a surveyor owns the property being sold or acts for a family member, a financially connected person or a close business associate.

The RICS requires that:

- Regulated Firms and Members must not accept an instruction where there is an actual or perceived conflict of interest unless all concerned parties provide written informed consent;
- Regulated Firms must implement procedures to manage conflicts of interest, including written records of decisions made and controls put in place.

Avoiding conflicts of interest is the preferred course of action. If a conflict of interest is identified, the Firm or Member should consider whether the instruction should simply be declined or if the conflict can be managed by seeking informed written consent and/or implementing an information barrier.

An information barrier, previously known as a Chinese Wall, requires the 'physical and/or electronic separation of individuals (or groups of individuals) within the same firm which prevents confidential information passing between them' (RICS, 2017). This could include separating administration teams and filing systems or handling the instructions in separate offices and by different surveyors. Particularly in small firms, setting up a genuinely robust information barrier may be impossible, so declining an instruction may be the most ethical course of action.

RICS also addresses the risk of conflicts of interest where Members and Regulated Firms act on open market sales or disposals of UK commercial property investments in the RICS Professional Standard Conflicts of Interest – UK Commercial Property Market Investment Agency (1st Edition).

Under no circumstances can a surveyor act for the buyer and the seller of the same property, known as dual agency (or double dipping).

Multiple introductions of an investment opportunity may be made (i.e., sending the details to more than one potential purchaser). However, the Terms of Engagement must confirm whether the surveyor acts on an exclusive or non-exclusive basis. If a non-exclusive basis is agreed upon, informed consent must be obtained from all clients, and appropriate information barriers must be put in place. Either way, the seller and selling agent must be made aware of the matter.

Incremental advice, e.g. a surveyor acting for the seller on a disposal but then being approached separately by the buyer or lender to provide a valuation, is permitted, providing that information barriers and informed consent are implemented.

Money Laundering, Bribery, Gifts and Hospitality

RICS Regulated Firms must comply with the mandatory RICS requirements relating to anti-bribery, corruption and money laundering. These are set out in the Professional Standard Countering Bribery and

Corruption, Money Laundering and Terrorist Financing (1st Edition, 2019) and the latest anti-money laundering regulations. This requires the firm to assess money laundering risks and put processes and procedures in place to carry out due diligence on clients, amongst other requirements.

The RICS Professional Standard sets out the following definitions (RICS, 2019):

- Bribery – The offer, promise, giving, demanding or acceptance of an advantage as an inducement for an action that is illegal, unethical or a breach of trust;
- Corruption – The misuse of public office or power for private gain or abuse of private power concerning business practice and performance;
- Customer Due Diligence (CDD)/Know Your Customer (KYC) – Taking the appropriate steps to ascertain who the customer or client is and, if relevant, their ultimate beneficial owner and any counterparty (i.e., other involved parties). These can be relatively simple checks to verify the identity of the customer/client or may entail deeper investigations. This is a legal and regulatory requirement in many countries;
- 'Facilitation payment' – A payment made to a government official to speed up a routine administrative action. Such payments are customary and legal in some countries, but in many jurisdictions, they are criminalised;
- Money laundering – Concealing the source of the proceeds of criminal activity to disguise their illegal origin. This may occur through hiding, transferring and/or recycling illicit money or other currency through one or more transactions or converting criminal proceeds into seemingly legitimate property;
- Red flags – Common characteristics that either individually or in combination might indicate potential misuse of the real estate sector for money laundering or terrorist financing purposes.

Money laundering is also defined by Part 7 of the Proceeds of Crime Act (POCA) (2002) as 'the process by which the proceeds of crime are converted into assets which appear to have a legitimate origin, so that they can be retained permanently or recycled into further criminal enterprises'.

Under the Money Laundering and Terrorist Financing (Amendment) Regulations 2022, a surveyor must undertake due diligence on clients (including the beneficial owner of the client company) before accepting instructions. This includes checks on both parties to a transaction, i.e., the purchaser and the seller in an acquisition or disposal. We cover the requirements of customer due diligence (CDD) later and in the Client Care section of this book.

Other key legislation includes the Terrorism Act 2000, where individuals must report knowledge or suspicion of terrorist financing to the National Crime Agency under POCA 2002. This created additional offences of:

- Concealing, disguising, converting or transferring criminal property;
- Becoming involved in an arrangement relating to criminal property;
- Acquisition, use or possession of criminal property;
- Disclosing/tipping off that a Suspicious Activity Report (SAR) has been submitted.

The Criminal Finances Act 2017 amended the preceding two Acts to provide law enforcement agencies with further powers of investigation. Penalties include unlimited fines and/or up to 14 years' imprisonment.

The Sanctions and Anti-Money Laundering Act 2018 established the UK Sanctions List. An up-to-date copy must be checked before a surveyor accepts new work from a client. Firms at higher risk (e.g., dealing with overseas clients or high-value or complex transactions) may use specialist software to ensure comprehensive due diligence is undertaken.

Suppose a red flag (i.e., a suspicious transaction) or a money laundering concern is identified. In that case, e.g., a breach of financial sanctions or frozen assets are held, a report must be submitted to the Office for Financial Sanctions Implementation (OFSI).

The Economic Crime (Transparency and Enforcement) Act 2022 established a Register of Overseas Entities for overseas entities holding UK real estate.

The Bribery Act 2010 applies to UK businesses operating in the UK and overseas and non-UK businesses operating in the UK.

A bribe is an 'exchange of something of value in return for someone doing or agreeing to do something improper in a business context' (RICS, 2019), e.g., holiday, money, contract, dinner, event tickets.

Whether a gift or hospitality is improper is based on the test of what a reasonable person in the UK would expect concerning the performance of that function or activity. The four offences under the Act are:

- To offer, promise or give a bribe;
- To request, agree to receive or accept a bribe;
- To directly or indirectly offer, promise or give something of value (e.g., facilitation payments) to a foreign public official to influence their decision-making, unless there is a written law permitting the practice;
- Failure to prevent a bribe.

There are six fundamental principles to be aware of under the Bribery Act 2010:

- Proportionality;
- Top-level commitment;
- Risk assessment;
- Due diligence;
- Communication;
- Monitoring and review.

It is a corporate criminal offence to fail to prevent bribery, and a senior officer of a corporate body could be liable for offences committed by others. There may, however, be a defence in court if adequate procedures are in place to prevent bribery.

Penalties under the Act include up to ten years' imprisonment and/ or an unlimited fine for individuals. Directors can be disqualified from acting for significant periods, whilst firms may face an unlimited fine.

Gifts and hospitality are common ethical dilemmas that surveyors may face. It is simply not a case of saying that you would decline every offer in your APC interview.

If the offer of a gift or hospitality is genuinely and proportionately for business purposes, then it may be reasonable to accept. The Ministry of Justice Guide to the Bribery Act 2010 (Ministry of Justice, 2010) confirms that

> you can continue to provide tickets to sporting events, take clients to dinner, offer gifts to clients as a reflection of your good relations or pay for reasonable travel expenses in order to demonstrate your goods or services to clients if that is reasonable and proportionate for your business.

Of course, if you consider that an offer could (or could be perceived to) affect your impartiality or integrity, you should politely decline it. In these circumstances, the offer could amount to a bribe if it was intended to induce you to behave without good faith, impartiality or contrary to your position of trust. You may face cultural differences if you work outside the UK, so this is something else to be aware of.

When considering whether to accept the offer of a gift or hospitality, you should consider the following:

- Is the value of the offer proportionate to the work undertaken?
- Who is the intended recipient; is it an individual or a range of people, such as a project team?
- Is the offer reasonable and proportionate?
- Is the offer for genuine business purposes?
- When was the offer made? Was it near a critical business decision or contract award (caution required!) or at a regular business marketing event?
- How easy is it to politely decline the offer?

Finally, turning to the RICS guidance on the aforementioned issues, the RICS Professional Standard Countering Bribery and Corruption, Money Laundering and Terrorist Financing (1st Edition) is split into three parts:

- Part 1 – Mandatory requirements;
- Part 2 – Good practice guidance;
- Part 3 – Supplementary guidance on Parts 1 and 2.

Part 1 is further split into sections:

- Bribery and corruption;
- Money laundering and terrorist financing.

Each section explains what RICS Regulated Firms and RICS members must do.

For example, regarding bribery and corruption, RICS Regulated Firms must not offer or accept bribes, must have relevant plans and policies, and must report any suspicious activity in line with the Bribery Act 2010. Firms should record their risk evaluation in writing, including the required due diligence. RICS requires full records evidencing compliance with the Professional Standard.

RICS members must also not offer or accept bribes, must have adequate knowledge of the Bribery Act 2010 and the Professional Standard, and must report any suspicious activity.

In relation to money laundering and terrorist financing, RICS Regulated Firms must not be involved in or facilitate any suspicious activity. As with bribery, appropriate systems, training and policies must be in place, and suspicious activity must be reported, as appropriate. Written records must be kept, and adequate due diligence must be conducted on clients and transactions. Similar requirements are placed on RICS members as for bribery.

Part 2 sets out what RICS Regulated Firms and members should do (rather than what they must do) relating to best practice. It is again split into two sections, as per Part 1.

In relation to bribery and corruption, RICS Regulated Firms should have a written and regularly updated anti-bribery and corruption policy, encourage transparency using a gifts and hospitality register, provide guidance and training to staff (including a code of behaviour), appoint a Responsible Person and/or champion within the firm and carry out due diligence on suppliers to ensure they are acting responsibly.

RICS members should use their firm's gifts and hospitality register appropriately, attend training and adhere to their firm's policies.

In relation to money laundering and terrorist financing, RICS Regulated Firms should again have written policies, provide staff training, identify the beneficial owner (i.e., the person or persons who ultimately own or control the company) of client companies and appoint a Responsible Person within the firm. Members have similar responsibilities as for bribery, including an awareness of legislation and complying with company policy.

Part 3 provides additional guidance on due diligence and taking a risk-based approach. The latter introduces the 3 Ws:

- Who you act for;
- What you are doing;
- Why you are being asked to do something.

The appendices to the Professional Standard include CDD, AML and reliance templates.

Terms of Engagement

When agreeing a new instruction, surveyors should always formalise the engagement in formal Terms of Engagement. These should be signed physically or electronically by both parties and held on file for future reference.

These should consider the following, amongst other key terms:

- Is the surveyor qualified, competent, skilled and experienced?
- Does the firm have appropriate PII for the work to be undertaken?
- Are there any conflicts of interest?
- Has sufficient client due diligence been undertaken, including money laundering checks?
- Is the fee basis set out clearly and transparently?
- Is the deadline for the work achievable and stated clearly?
- What are the payment terms?
- Have any referral fees been declared?

Complaints Handling Procedure (CHP)

All Regulated Firms must operate a CHP as per the Rules of Conduct. Firms must also operate a complaints log, nominate a Complaints Handling Officer and include an approved redress mechanism within their CHP.

A firm's CHP should be in writing and split into two stages; the first providing internal consideration of the complaint and the second providing independent redress if the firm cannot resolve the complaint.

Clients must be made aware of the CHP in the firm's Terms of Engagement, and a copy of the CHP must be provided to a client if a complaint is received. If a complaint is received, the firm should also provide notice of this to their PII provider.

Additional guidance on complaints handling, published initially for residential surveyors and valuers, can be found in the RICS Guidance Note Complaints Handling (1st Edition).

Professional Indemnity Insurance (PII)

RICS Regulated Firms must have PII cover in place. PII protects firms from negligence claims made by clients unsatisfied with the professional services provided to them. It may also cover a firm for loss of documents or data, intellectual property claims and defamation. PII protects clients from financial losses that a firm cannot meet.

The RICS requires that PII must meet the following minimum requirements (RICS, 2022b):

- Provide cover for any one claim or be on an aggregate plus unlimited round-the-clock reinstatement basis;
- Fully retroactive ('each and every claim made' basis), covering claims made during the term of insurance cover irrespective of when the original act occurred;
- Include RICS' minimum policy wording. In 2022, this included specific fire safety exclusions;
- Being written on a full civil liability basis;
- Provide at least the minimum level of indemnity based on the firm's turnover in the previous year (or estimated for a new firm);
- Provide for a maximum level of uninsured excess based on the sum insured;
- Underwritten by an RICS-approved insurer;
- Provide cover for past and present employees.

As of 2023, the RICS specifies minimum limits of indemnity, based on the last year's turnover:

- £100,000 or less (last year's turnover) = £250,000 (minimum limit of indemnity);
- £100,001 to £200,000 = £500,000;
- £200,001 and above = £1,000,000.

As of 2023, RICS also specifies maximum levels of uninsured excess, based on last year's turnover:

- £10,000,000 or less (last year's turnover) = greater of 2.5% of the sum insured or £10,000 (maximum uninsured excess);
- £10,000,001 and above = no limit.

Candidates should always check the current limits and requirements of RICS concerning PII.

If a firm cannot obtain adequate PII (or run-off cover), it can apply to use the RICS Assigned Risks Pool (ARP).

Suppose a firm receives a claim (i.e., a letter notifying the firm of legal action). In that case, the threat of a claim or encounters of circumstance which may lead to a claim or a complaint via the firm's CHP, then the firm's PII insurer must be notified as soon as possible.

When an RICS Regulated Firm or Member (sole trader) stops trading, they must have a minimum of £1,000,000 of run-off PII cover. This is required for a minimum of six years for consumer claims, although a more extended period may be required in certain circumstances. For commercial claims, adequate and appropriate cover is necessary for at least six years.

Having run-off cover is an essential part of risk management for closed firms, particularly given the case of *Merrett v Babb* [2001] QB 1174. In this case, a valuation surveyor conducted a mortgage valuation for a lender. His firm was subsequently closed due to bankruptcy with no run-off cover. The house purchaser successfully brought a negligence claim against the surveyor personally, even though the purchaser had not seen the valuation report and was unaware of the surveyor's identity when they purchased the property.

Essentially, following the case of *Hedley Byrne & Co Ltd v Heller & Partners Ltd* [1964] AC 465, a duty of care is established so where run-off cover is not in place, a surveyor could be held personally liable for negligence relating to work undertaken in a former company.

The RICS has published the Guidance Note Risk, Liability and Insurance (1st Edition) to provide guidance on mitigating or avoiding the risks and liabilities associated with surveying work. Surveyors may face claims under breach of contract and negligence.

When providing surveying services, a surveyor will agree a contract (or Terms of Engagement) with the client defining each party's scope of service and obligations. The client (i.e., the other party to the contract) can take action against the surveyor if any of the terms are breached. There is an implied term that the surveyor will provide services with reasonable skill and care, i.e., not to act negligently.

A claim can also be brought under the tort of negligence. A tort is a civil wrong recognised by law rather than a breach of contract. This requires the same standard of service as the implied term discussed earlier, i.e., to act as a reasonable professional would do in the circumstances.

However, a claim under the tort of negligence can be brought by anyone (i.e., a third party) to whom the surveyor owes (or has assumed) a duty of care. This is wider than a breach of contract which can only be brought by the contracting party, i.e., the client. When considering whether a duty of care is owed to a third party, the court will consider whether the damage suffered was reasonably foreseeable as a result of the defendant's actions, whether the relationship between the parties was sufficiently proximal and whether it is just, fair and reasonable to impose a duty of care.

This duty of care could arise when a surveyor allows a third party to rely upon their advice. This means that the third party can acquire similar rights to a client's but without the protection of the surveyor's contract and any liability cap.

Specific instances where a third party may be owed a duty of care include owner-occupier purchasers (borrowers) in mortgage valuations, following the cases of *Smith v Bush* [1990] 1 AC 831 and *Harris v Wyre Forest District Council* [1988] QB 835. This will not extend to buy to let purchasers (borrowers), however, following *Scullion v Bank of Scotland Plc* (*t/a Colleys*) [2011] EWCA Civ 693.

Surveyors should, therefore, be very careful when defining in their Terms of Engagement what the client may use the surveyor's work for. Ideally, this should only be for the sole purpose intended. If a client then relies on the advice for any other purpose, the Terms of Engagement should exclude liability for losses incurred.

Damages (for financial loss) are generally awarded for both to put the complainant in the position that they would have been if the contract were not breached or the negligent act not committed, i.e., no better or worse off.

Under contract, the loss must have been in reasonable contemplation when the contract was agreed. Under tort, both parties must have reasonably foreseen the loss when the negligent act occurred. Any other losses will generally not be compensated for.

Generally, claims are made against the firm who provided the services. However, claims can be brought against individual partners in a partnership. This promotes the benefits of using company structures such as an LLC or LLP, where the partners' personal liability is limited. To avoid any problems, all Terms of Engagement should expressly exclude personal liability.

As Professional Indemnity Insurance (PII) is agreed on a 'claims made' basis, the policy must be in place when the claim is made (irrespective of when the services were provided). This emphasises the importance of having run-off cover if a firm ceases to trade for a sufficient period.

A liability cap limits the amount of damages that can be claimed by a client in the event of loss, even in the event of a negligent act. This helps firms to manage risk around the provision of surveying services. Liability caps must be written explicitly into a firm's Terms of Engagement and must be reasonable to be enforceable. What reasonable means, however, may differ between a business or a consumer client, and firms must consider this when agreeing to liability caps.

Reasonableness relates to:

- Risk level;
- Level of fee;
- PII limit and premium;
- Potential liability without a cap;
- Type of client;
- Contractual wording, ideally noted in the contract and any covering engagement letter.

Example wording for a liability cap could be (RICS, 2022a):

> our aggregate liability arising out of, or in connection with, these services, whether arising from negligence, breach of contract, or any other cause whatsoever, shall in no event exceed £[x]. This clause shall not exclude or limit our liability for actual fraud and shall not limit our liability for death or personal injury caused by our negligence.

There is also a difference between liability caps and a firm's PII limit. The former is the contractual agreement between the client and the firm, i.e., the maximum amount that would be paid to the client in the event of a claim. The maximum amount the insurer will pay out for a claim is the latter. It is best practice, therefore, to agree liability caps below the PII limit.

A liability cap could be set as a multiple of a proposed fee or a percentage of a value reported, amount loaned or value of a construction contract.

In summary, to appropriately manage risk, the Terms of Engagement should include:

- Scope of work;
- Fee basis;

- Liability cap;
- Contracting entity (i.e., the firm) and exclusion of personal liability;
- Proportionate liability if advice is provided alongside other professionals, such as a solicitor or architect;
- Third-party reliance;
- Alternative dispute resolution method to avoid costly court litigation;
- Governing law and jurisdiction, i.e., requiring any claims to be subject to English law (or the firm's home country).

Client Money

Many Regulated Firms will handle clients' money during their professional work. Any monies held must be handled appropriately via robust controls and systems.

RICS provides requirements for client money handling in the Professional Standard Client Money Handling (1st Edition).

It is a fundamental obligation for RICS Regulated Firms to keep clients' funds separate from those of the firm. This is a duty that can be discharged by maintaining proper accounting records.

Client money is defined by RICS in the Professional Standard (RICS, 2022c) as any cash, cheque, draft or electronic transfer which an 'RICS Regulated Firm holds for or receives on behalf of another person, including money held by a Regulated Firm as stakeholder and is not immediately due and payable on demand to the RICS Regulated Firm for its own account'. Examples include rent, service charge or interest credited to a client account.

RICS (RICS, 2022c) defines client money to exclude

'fees paid in advance for professional work agreed to be performed, and identifiable as such, unless the fees are for work undertaken as a property agent as defined by the Rules of the RICS Client Money Protection Scheme for Property Agents'.

Office money could include fees, disbursements and interest on general client accounts, which the client agrees (in writing) will not accrue to them.

Although not an exhaustive list, some of the essential requirements include:

- Not holding office money in a client money account;
- Ensuring that the word 'client' is in the client account name;

- Holding all funds in a client money account over which the firm has exclusive control;
- Ensuring that client money is immediately available;
- Confirming the bank operating conditions in writing;
- Keeping comprehensive and accurate accounting records, systems and procedures. This includes regular reconciliations by a senior staff member or Principal, having disaster recovery plans in place, ensuring electronic data is held securely and avoiding overdrawn balances.

When client money is received, RICS Regulated Firms must promptly pay it into the specific client account. If mixed funds are received, they must be paid into the client account, and office money must be transferred out promptly.

Unidentified funds should be promptly identified. If the owner is not identified within three years, they must be paid to a registered charity.

For payments out of a client account, RICS Regulated Firms must use the money only for the client's matters and with the client's written consent. Client money should be returned as soon as it is no longer needed. Firms should ensure sufficient funds are held before payments are made and obtain written consent for direct debits, standing orders or bank charges.

Candidates should review their firm's procedures for holding client money in light of the requirements of the RICS Professional Standard.

Any firms holding client money must also pay an annual regulatory review fee following the completion of their Annual Return. This contributes to the operation of the RICS Client Money Protection Scheme, which provides free recourse (within 12 months of discovering the loss) for clients when a Regulated Firm is unable to repay a client's money, subject to specific limits (i.e., £50,000 per claim) and exceptions. The Scheme is one of last resort, i.e., it will reimburse the client for a direct loss of funds where the firm cannot make good the loss.

Claims may be made in circumstances such as a firm's insolvency, the death of a sole practitioner, misappropriation of client monies or transfer of client monies to another organisation.

The Scheme is split into two categories:

- Client Money Protection for Surveying Services;
- Residential agency under the Client Money Protection Schemes for Property Agents (Approval and Designation of Schemes) Regulations 2018 (Client Money Protection for Residential Agents).

RICS Regulated Firms

After qualifying as a Chartered Surveyor, some candidates will decide to pursue self-employment or to set up their RICS Regulated Firm. This requires knowledge and application of many essential RICS requirements, including holding adequate insurance, a CPD policy and a Complaints Handling Procedure. Setting up a new practice is not something to be undertaken lightly. It requires robust knowledge of the legal and regulatory requirements and good business and financial sense.

There are many benefits of being regulated by RICS, including:

- Committing to providing the highest standards of service and professionalism;
- Being open to independent scrutiny;
- Taking an ethical approach to business;
- Being protected via PII;
- Providing security, comfort and redress for clients by having a CHP;
- Providing a guarantee to undertake continuing professional training and having the relevant experience to undertake professional work.

It is free to register a firm for regulation by RICS. The process is simple and requires a firm Details Form to be filled out and submitted to RICS online.

The following details are required:

- Nomination of a Contact Officer – an RICS Member who will be the main point of contact with RICS;
- Nomination of a Responsible Principal – an RICS Member who will ensure compliance with RICS Regulations and standards;
- Firm name, contact details, type and registration numbers;
- Details of all RICS Principals (partners or directors) and employees.

After submitting the Firm Details Form, the firm must provide further registration details using the RICS Firms Portal.

Firms wishing to register for regulation must meet the following requirements, set out in the RICS Rules for the Registration of Firms (Version 8), effective February 2022:

- The firm must offer professional surveying services to the public (clients);
- The firm must operate in Regulated Area A, i.e., the UK;

- The firm must have at least 25% of the Principals who are AssocRICS, MRICS or FRICS qualified.
- If a firm has over 50% of Principals who are AssocRICS, MRICS or FRICS, then it must register for regulation as a mandatory requirement;
- The firm must comply with the RICS Rules of Conduct at all times;
- Firms must designate a Responsible Principal who provides oversight, accountability and engagement with RICS standards and Regulations.

Once a firm is regulated by RICS, they can use the RICS logo and the designation 'Chartered Surveyors' within their trading name. In the Rules for the Use of the RICS Logo and Designation by Firms (Version 6), effective February 2022, it is a mandatory requirement to state that the firm is regulated by RICS on all business material. The firm can also be listed in the RICS Find a Surveyor directory. It is also a requirement that Regulated Firms must include prescribed text in their Terms of Engagement issued to clients in respect of surveying services to explain what being regulated by RICS means.

Sole practitioners or sole directors in a corporate practice should have in place arrangements in the event of death, incapacity or other extended absences (e.g., holiday or sabbatical) from the business. This is generally achieved through having a locum, another professional appointed to 'stand in' for the surveyor if they cannot work. A sole practitioner may appoint a locum as their complaints handling officer to ensure their Complaints Handling Procedure is run fairly and impartially.

Typically, a locum will be another Chartered Surveyor, although they could also be a solicitor or accountant by trade, i.e. a trusted professional.

A locum should be appointed formally in writing, documenting the following:

- Appointment details;
- Confidentiality clauses;
- Any conflicts of interest;
- Fee basis;
- Termination;
- Professional Indemnity Insurance (PII) requirements;
- Process for the locum to act independently or if they need to seek instructions before proceeding.

Locums must be covered for any work undertaken whilst a sole practitioner is unavailable. This could be under their cover or the firm's policy if appropriate. If a locum runs a firm for any period, then RICS must be notified.

If a sole practitioner were to die, their PII cover would need to be converted to run-off cover. This would relate to work undertaken by the sole practitioner and any former employees and would also help to protect their estate.

A locum may need to handle client money during a sole practitioner's absence. This means they will need to be able to access relevant accounts in line with a signed agreement with the bank. They will also need to be familiar with and act in line with the RICS requirements for client money handling.

They may also need to access client files to ensure ongoing work can continue. This means they will need access to electronic and hard copy files, including knowing passwords and having a list of key contacts. They must also be familiar with legal, banking, tax and accounting processes.

If the sole practitioner cannot return to work, the locum may also become responsible for overseeing the sale of the firm.

In addition to the preceding RICS requirements, a Regulated Firm may need to adhere to many statutory requirements. It should be noted that the following list is not exhaustive but provides a starting point for any newly Regulated Firm:

- Business Names Act 1985 and Companies Act 2006;
- Health and Safety at Work Act 1974;
- Bribery Act 2010;
- Anti-money laundering legislation;
- Data Protection Act 2018;
- Employment law.

Firms must also consider taxation requirements, e.g., VAT and Corporation Tax. Insurance must also be considered, including public liability, employer's liability, building, contents, cyber and business interruption.

Surveyors also need to be aware of what would happen in the event of the closure of an RICS Regulated Firm. This includes:

- Informing RICS;
- Informing clients at the earliest opportunity and, if required, confirming hand-over arrangements to a new firm;
- Returning clients' monies;

- Informing the firm's PII insurer and arranging run-off cover;
- Retaining records for at least six years.

Phoenix firms do not close in an orderly manner, typically because of insolvency or not having PII run-off cover. Substantial risk is posed to the public interest when these firms rapidly re-establish themselves as new trading entities.

How Can You Demonstrate Levels 1, 2 and 3?

Your assessors will start at the highest level you have declared, which will be level 3. Taking ethics to the level of giving reasoned advice, these are examples of relevant scenarios of giving ethical advice or advising on ethical dilemmas that you might encounter in your every-day working practice.

- Dealing with a complaint in line with the firm's CHP;
- Handling client's money appropriately and in line with relevant RICS guidance;
- Dealing with PII cover or claims;
- Registering a firm with RICS;
- Advising on a conflict of interest and appropriate action to resolve an issue.

Some example questions include:

- In your submission, you gave an example of a conflict of interest. Talk me through the advice you gave to manage this;
- Your submission gave an example of being offered a gift during a tender. How did you deal with this ethically?
- When asked by your client to increase your valuation figure, explain the advice you gave on how to act ethically and professionally.

At level 2, examples of relevant work-based activities include:

- Being involved with RICS through a board, committee or other activity;
- Identifying a conflict of interest and taking appropriate action to manage it or to decline the instruction;
- Agreeing transparent and reasonable professional fees;

- Undertaking pre-instruction checks, including those for anti-money laundering;
- Agreeing Terms of Engagement when taking on new work;
- Dealing with an offer of a gift or corporate hospitality;
- Complying with RICS CPD requirements;
- Ensuring that an employer's processes comply with RICS requirements, including the Rules of Conduct.

Typical level 2 questions will start with, 'How have you . . . ?', 'Tell me about an example when you . . . ?' or 'How did you provide good client care when . . . ?'. Remember, all assessors' questions should be based upon the examples you provide in your level 2 write-up for this competency in your summary of experience – rather than being hypothetical or general.

Examples include:

- In your submission, you gave an example of agreeing a reasonable fee basis. Which Rule(s) of Conduct did you apply?
- What pre-instruction checks did you carry out at the valuation example in your submission?
- How did you check for conflicts of interest in the letting example in your submission?

You may also be asked some level 1 questions, focusing on the knowledge behind your practical level 3 (and potentially level 2) examples.

Be prepared to know the detailed requirements of the RICS relating to complaints handling, Terms of Engagement or money laundering checks, for example.

- What are the RICS Rules of Conduct?
- What RICS guidance do you comply with when agreeing a new instruction?
- What RICS guidance relates to bribery?

Conclusion

In conclusion, acting ethically is at the heart of what it means to be a Chartered Surveyor. It is not simply enough to be sufficiently competent, skilled and knowledgeable. This is evidenced by the requirement for all APC candidates to achieve level 3 in the Ethics, Rules of Conduct and Professionalism competency.

In the final assessment interview, an incorrect or unethical answer concerning this competency will constitute an automatic failure, irrespective of a candidate's performance in the rest of the interview. AssocRICS candidates must also demonstrate their strong understanding and application of ethics and professionalism through the RICS Professionalism Module.

Chartered Surveyors must seek to display the highest professional standards in their personal and professional lives. This should be an aspiration for all existing and future Chartered Surveyors, irrespective of sector, market or discipline. It is a mandatory, not a voluntary requirement, and never something to be taken lightly.

There will be times when a surveyor will need to take responsibility to rectify a problem or a mistake or address a challenge, and it is the desire and requirement to act ethically that will ensure that these situations are dealt with diligently and professionally. Undertaking the mandatory three-yearly CPD relating to this competency once qualified as AssocRICS or MRICS will support surveyors in making ethical, professional and moral decisions in times of uncertainty or difficulty.

Reference List

Forsyth, J. & Wilcox, J., 2022. *Real Estate: The Basics*. London: Routledge.

Harris v Wyre Forest District Council [1988] QB 835 (1988).

Hedley Byrne & Co Ltd v Heller & Partners Ltd [1964] AC 465 (1964).

Merrett v Babb [2001] QB 1174 (2001).

Ministry of Justice, 2010. *Guide to the Bribery Act 2010*. [Online] Available at: www.justice.gov.uk/downloads/legislation/bribery-act-2010-guidance.pdf [Accessed 13 July 2023].

Plimmer, F., Edwards, G. & Pottinger, G., 2009. *Ethics for Surveyors: An Educational Dimension. Commercial Real Estate Practice and Professional Ethics*. London: RICS.

Proceeds of Crime Act (2002).

RICS, 2017. *Professional Standard Conflicts of Interest*. 1st ed. London: RICS.

RICS, 2019. *Professional Standard Countering Bribery and Corruption, Money Laundering and Terrorist Financing*. 1st ed. London: RICS.

RICS, 2020a. *Governance Procedures and Processes*. [Online] Available at: www.rics.org/globalassets/rics-website/media/governance/standing-orders/ [Accessed 5 October 2020].

RICS, 2020b. *RICS Bye-Laws*. [Online] Available at: www.rics.org/globalassets/rics-website/media/governance/bye-laws/ [Accessed 5 October 2020].

RICS, 2020c. *RICS Regulations*. [Online] Available at: www.rics.org/globalassets/rics-website/media/upholding-professional-standards/regulation/regulations/ [Accessed 5 October 2020].

RICS, 2020d. *Royal Charter*. [Online] Available at: www.rics.org/globalassets/rics-website/media/governance/royal-charter/ [Accessed 5 October 2020].

RICS, 2022a. *Guidance Note Risk, Liability and Insurance*. [Online] Available at: www.rics.org/content/dam/ricsglobal/documents/standards/risk_liability_and_insurance_1st_edition_1.pdf [Accessed 13 July 2023].

RICS, 2022b. *Professional Indemnity Insurance Requirements*. [Online] Available at: www.rics.org/content/dam/ricsglobal/documents/standards/April_2022_Professional_Indemnity_Insurance_Requirements_Version_9.pdf [Accessed 13 July 2023].

RICS, 2022c. *Professional Standard Client Money Handling*. [Online] Available at: www.rics.org/content/dam/ricsglobal/documents/standards/Client%20money%20handling_Oct22.pdf [Accessed 13 July 2022].

Scullion v Bank of Scotland Plc (t/a Colleys) [2011] EWCA Civ 693 (2011).

Smith v Bush [1990] 1 AC 831 (1990).

Chapter 3

Client Care

Client Care is a mandatory RICS APC competency required to level 2. This means that you must write up your summary of experience, including a statement at levels 1 and 2. Some pathways allow you to take Client Care as a technical competency to a higher level, i.e., level 3.

This competency should be read in tandem with Ethics, Rules of Conduct and Professionalism, which covers many other topics relevant to Client Care, such as conflicts of interest and pre-instruction checks.

The term 'client' does not just cover external (contractual) clients, for example, where a surveyor in a consultancy firm is acting for a landlord or a tenant client. It also covers internal clients, such as shareholders, a board or a line manager, and other stakeholders to whom the surveyor owes a duty of care.

The RICS pathway guide sets out the relevant knowledge (level 1) and practical application (level 2) for this competency. However, you do not need to have encountered everything listed, as this will depend on your experience and role.

This chapter will explain what you should know concerning Client Care, covering the main requirements of level 1.

Client Care Skills

Rule 3 of the Rules of Conduct confirms that Members and Regulated Firms must provide good-quality and diligent services.

A crucial part of achieving this is through defining a clear client brief. The initial point of contact with a client could be through an enquiry email or telephone call. At this stage, the surveyor must keep detailed notes on the client's requirements before following up with a more substantial meeting, virtually or in person.

DOI: 10.1201/9781003437116-3

Clients often do not know what they need, so the surveyor should be curious and ask intelligent questions to establish the problem or requirement and advise on a service or solution.

The client's brief can then be defined based on the surveyor's initial due diligence, discussions and desktop research. Ideally, the brief should be recorded in writing and followed with clear Terms of Engagement.

A surveyor's initial research and due diligence must include research on the client, context and property or project. This will form part of the surveyor's file and be used during the instruction when reporting to and advising the client.

A client brief must be tailored to the client and the instruction in question. Key things to include (not an exhaustive list) are:

- Date;
- Surveyor's name;
- Client name;
- Statement of the problem or instruction;
- Aims;
- Proposed solution or service;
- Pricing;
- Deadlines.

A client brief may need to include several proposed solutions to a problem or instruction. This could be through different service levels or an options appraisal. In all cases, a surveyor should avoid advising the client on the solution until the Terms of Engagement are agreed in writing (at this point, a Firm's Professional Indemnity Insurance (PII) policy will cover the surveyor for advice given).

The Terms of Engagement will then confirm the full details of the instruction.

Defining the client's brief and the scope of service is an essential part of the client care process. The scope of service needs to be within the surveyor's competence, skills and experience. A surveyor may, for example, have undertaken similar work before and have a track record of proven results. Equally, a surveyor may need to seek internal or external assistance to complete the instruction to the required level. If external specialist advice is necessary, the surveyor must deal with sub-contracting in their Terms of Engagement or ensure their client instructs the third party directly. It is possible to work in new practice areas, but this will require assistance from a more experienced surveyor and additional relevant training.

A surveyor must also act within their firm's PII limits (see the Ethics, Rules of Conduct and Professionalism chapter). If in doubt, a surveyor should always check their PII policy or speak to their firm's broker. Suppose a surveyor acts outside the scope of their firm's PII policy. In that case, they expose themselves and their firm to the risk of an uninsured claim and significant financial liability. Other insurance required to protect a surveyor and their firm is covered in the Ethics, Rules of Conduct and Professionalism competency.

A surveyor must also agree fees with the client at this stage. Fees can be calculated in various ways, e.g. fixed fee, hourly rate, performance-based fee (although not when acting as an expert witness at a tribunal, third party proceedings or court) or a percentage fee of the instruction's outcome. In all cases, the fee should be proportionate to the work undertaken and the surveyor should be able to justify the fee proposed. This may require the surveyor to calculate the time necessary to complete the instruction, which can be challenging as time requirements vary.

If a client requests that a fee proposal is reduced, the surveyor should avoid aggressive undercutting or reducing a fee without an accompanying reduction in service scope.

Surveyors could have a price list for set services to help client decision-making. However, they should ensure this sets out what is included and excluded. This may only be appropriate for specific surveyors and services, e.g. a residential surveyor providing level 1, 2 and 3 surveys or a fixed percentage agency fee.

Furthermore, the surveyor must be transparent with the client if referral fees are paid to another party. These could be a referral fee paid to another firm or professional for recommending the surveyor or firm to undertake the work.

A surveyor should ensure they agree realistic reporting deadlines with the client. This should consider existing workload and ensure the surveyor can provide a high-quality service within the scope of service.

A wide variety of behaviours allow a surveyor to provide a high standard of client care. Some of these are outlined as follows:

- Clear communication using agreed methods;
- Account management – ensure regular contact away from live instructions to meet long-term business objectives (holistic approach);

- Feedback, e.g. project reviews, satisfaction surveys, discussion groups and quality assurance procedures;
- Business development activities and events;
- Understanding client objectives and requirements;
- Active listening;
- Responding promptly to queries;
- Regular reporting – discuss at the outset and raise potential delays or issues in advance (and ideally over the telephone first);
- Management of client data, including clear records of all communication (both written and verbal);
- Avoid technical jargon for lay clients. Surveyors could send out an appropriate RICS Consumer Guide to explain complex matters;
- Consider cultural differences and translation, if appropriate;
- Consider neurodiversity, disability and diverse user needs, including braille, large print or accessible document formats.

Surveyors must also ensure they are accurate, transparent and honest when advertising their services.

They must consider the Consumer Protection from Unfair Trading Regulations 2008 (CPRs) and Business Protection from Misleading Marketing Regulations 2008 (BPRs).

This means that they must:

- Give sufficient information;
- Avoid misleading information.

And that they must not:

- Act aggressively;
- Engage in banned practices;
- Make unfair comparisons with competitors.

Surveyors who provide services to the public (consumers) must be aware of the Consumer Rights Act 2015, requiring them to be transparent regarding fees and VAT.

The Consumer Contracts (Information, Cancellation and Additional Charges) Regulations 2013 provides consumers with the legal right to a 14-day cooling off period if the Terms of Engagement were not signed in your office. The cooling-off period does not apply if the client signs the Terms of Engagement at your premises.

This right should be referred to within the Terms of Engagement, and work should not commence until the cooling-off period has expired.

However, suppose you agree to commence work immediately. In that case, you should state this in the Terms of Engagement, alongside a statement confirming that your client will be liable for any costs incurred within the initial 14-day period (including if the contract is cancelled within this period).

A good advertising or marketing campaign should take a consistent multi-channel approach. Using case studies and testimonials as part of this is advisable, although a surveyor must ensure they have their client's consent to use these. Surveyors can also request online reviews, such as on Trust Pilot or Google as part of the client feedback process. However, there is always the risk of negative reviews, and a surveyor needs to consider how they would respond respectfully and with integrity.

Confirming Instructions

Before accepting client instructions, a surveyor should undertake anti-money laundering checks. The key legislation is the Money Laundering and Terrorist Financing (Amendment) (No. 2) Regulations 2022.

Customer Due Diligence (CDD) means that the surveyor should check that the client is who they say they are.

Checks need to be undertaken when a new business relationship is established when the circumstances of an existing client change or for occasional transactions over EUR 15,000 (or EUR 10,000 for high-value dealers). Estate and letting agents must conduct due diligence on both parties to a transaction, not just their client. If they fall under the definition of carrying out estate agency activity under Section 1 of the Estate Agents Act 1979, they must also register with HMRC for anti-money laundering purposes.

Letting agents of commercial and residential property must carry out client due diligence (CDD) checks when a property is let for a term of a month or more, and at a rent, which, during at least part of the term is, or is equivalent to, a monthly rent of over EUR 10,000.

CDD includes:

- Checking the client's name, photograph, residential address and date of birth, using a passport, driving licence, bank statement and/ or utility bill;
- Obtaining information on the client company, e.g. from Companies House.

The surveyor may also need to identify the 'beneficial owner'. This is the individual who owns or controls the client company, or the party on whose behalf the instruction is carried out. This could include investigating the ownership structure of a company, partnership or trust.

Money laundering due diligence also includes checking the purpose of the business relationship and transaction, details of the client's business and the source of any funds.

If a surveyor doubts the client's identity, they should not act on their behalf until further investigations have been made. Surveyors should also monitor client companies or activity and repeat due diligence checks if circumstances change or they doubt the client's identity, intentions or funding source.

Documentation must be kept on file for at least five years from the end of the business relationship or completion of the transaction.

Enhanced Due Diligence (EDD) is required where a transaction is higher risk, e.g. due to the customer not being physically present when carrying out ID checks, where there is a higher risk of money laundering or the client being a politically exposed person (PEP). Additional due diligence is required in these scenarios, such as further identity checks, credit checks or investigating the source of funds.

Larger firms may have compliance teams to assist with or complete these checks, although surveyors should know the process and what is required.

To formalise the instruction, a surveyor must agree clear Terms of Engagement. This can help to avoid conflict and complaints. The surveyor should speak to the client to ensure they understand what they are instructing and that the terms reflect the client's brief.

A comprehensive set of Terms of Engagement should include the following:

- Client identity;
- Surveyor responsible for the case, including their qualifications;
- Scope and nature of work;
- Limitations (see risk management and liability caps in the Ethics, Rules of Conduct and Professionalism competency) and third-party reliance;
- Data protection policy;
- Due diligence undertaken, including what will not be checked and what assumptions will be made;

- Deliverables, including timeframes for reporting;
- Fee basis;
- Reference to the firm's Complaints Handling Procedure.

If applicable, a surveyor may need to consider other RICS guidance, such as the RICS Valuation – Global Standards (Red Book) for valuation work.

Internal clients may not require Terms of Engagement. However, a surveyor who primarily advises internal clients must still understand when and why Terms of Engagement are necessary. They will also always need to check for conflicts of interest, which can arise internally and externally.

The RICS also has Standard Forms of Consultant's Appointment available. These are typically used on construction projects, with various forms available depending on the project size. These are all available to download on the RICS website.

Surveyors may be asked to complete a Pre-Qualification Questionnaire (PQQ) before being awarded a contract or instruction. This will include questions before tendering on experience, capacity and financial standing. This allows the client to draw up a shortlist of suppliers to be invited to tender.

A surveyor may also come across an Invitation to Tender (ITT), which will be issued to tendering parties and includes details of the project or instruction. This usually defines a standard format for tender submissions and the requirements for tender bids, e.g. deadlines. Tender requirements are typically strict, so a surveyor must ensure they read them carefully before responding.

Client Management

Surveyors must have a reliable system for recording client details and instructions. This could be through a bespoke or off-the-shelf Customer Relationship Management (CRM) to manage existing and potential client relations. In smaller firms, this could be a simple spreadsheet or word-processed document.

Surveyors must also consider new business generation, including new clients, existing clients or cross-selling within a business.

A surveyor may act on behalf of many different types of clients. They will all come with varying levels of property knowledge, objectives and strategies. Therefore, defining a clear client brief for each client is essential.

We have already touched upon the fact that the definition of a client can also include other stakeholders. The RICS provides guidance on stakeholder engagement in the RICS Guidance Note Stakeholder Engagement (1st Edition). This defines a stakeholder as, 'individuals or groups who have an interest in a project or instruction because they are involved or affected by the outcomes'.

Furthermore, stakeholder engagement is defined as the 'systematic identification, analysis, planning and implementation of actions designed to engage stakeholders' (APM, 2023).

RICS suggests the use of the following to manage key stakeholders:

- Mendelow's power-interest grid – used to categorised stakeholders;
- CASE six-step approach – to analyse stakeholder engagement.

Key Performance Indicators (KPIs) are an essential part of client management. They can be used to monitor performance, progress or costs. They provide a helpful client feedback mechanism, providing that they are reported on regularly and improvement measures implemented if under-performance is identified.

An example of a set of KPIs for a landlord and tenant surveyor acting for a tenant occupying multiple properties could be:

- Lease events reported at least six months in advance;
- Budget rents reviewed three monthly;
- Lease events settled within six months of the event date unless exceptional circumstances apply and are agreed upon with the client.

Quality management is also a crucial part of client management, with the approach depending on the size of the firm. A firm must ensure that processes are followed to manage risk and ensure consistency and that work is delivered competently and based on accurate information and analysis.

A Quality Management System (QMS) could be used in larger firms, alongside compliance with ISO 19001:2015 (following the process of plan, do, act and check). This is a set of interrelated policies, processes and procedures in the core areas that can impact a firm's ability to meet a client's requirements.

A suggested procedure within a QMS could be:

- Conflict of interest checks;
- Customer due diligence process;
- Critical task checklist for files;
- Data handling and security processes;
- Process to check the quality of work, especially for new or junior staff;
- Billing process;
- Complaints handling process, including feedback on lessons learnt.

File management is part of quality management. Good record-keeping is vital if a claim arises or a client instructs further work.

This could include:

- Notes of checks undertaken;
- Terms of engagement and evidence of the agreement;
- Copies of all documentation;
- Records of critical decisions;
- Evidence of research, analysis, reasoning and calculations to support advice;
- Copy of deliverables provided to the client;
- Internal and external correspondence, including notes of phone calls or meetings;
- Peer-review feedback;
- After-delivery work.

Complaints

A complaint can be defined as any expression of dissatisfaction. Complaints generally arise when expectations are unmet, e.g. timing, scope of work or misunderstanding. This is why having a clear client brief and Terms of Engagement is so important.

As such, Terms of Engagement should set clear expectations and include reference to a firm's Complaints Handling Procedure (CHP).

Initially, a surveyor should always try to avoid complaints by seeking feedback and ensuring effective communication with clients. This includes identifying early warning signs of issues, e.g. lack of communication or non-payment of invoices, and seeking to resolve the issue before a complaint arises.

Firms must have a written Complaints Handling Procedure (CHP), as required by the RICS Rules of Conduct. This must be agreed with their PII provider and include a complaints log.

An idea of how a CHP should operate is outlined next:

- The surveyor can deal with dissatisfaction, but serious issues will need to be referred to the Regulated Firm's CHP, particularly if a complaint is received in writing from a client;
- Complaint received;
- Consider whether the PII provider needs to be informed (check the policy);
- Recorded and receipt acknowledged (seven days);
- Passed to complaints handling officer who investigates;
- Correspondence with the client to outline issues and course of action (if you cannot give a complete response immediately, then an update should be issued within 28 days);
- Outcomes recorded and communicated to the client;
- Lessons learnt shared internally.

If the complaint is unresolved, ADR can be pursued using a mechanism approved by the RICS Standards & Regulation Board (Version 10, effective February 2022). For example:

- Consumer redress (customer complaints) – Centre for Effective Dispute Resolution (CEDR), The Property Ombudsman, The Property Redress Scheme, Financial Ombudsman Service;
- Business to business (B2B) redress (contractual disputes) – Centre for Effective Dispute Resolution (CEDR), RICS Dispute Resolution Service (DRS).

RICS publishes guidance for surveyors and firms on complaints handling in the Guidance Note Complaints Handling (1st Edition). Although this primarily focuses on residential surveyors and valuers, it provides helpful guidance to all disciplines. The guidance confirms that a client's reasonable expectations can be set by Terms of Engagement, e.g. scope, quality, timeliness and understanding of service/advice.

In a residential RICS Home Survey, the surveyor has a duty to check that a contract is signed, that the client is choosing an appropriate level of survey and that the client clearly understands the service to be provided. The guidance also emphasises that it is always essential to understand the complaint before seeking to resolve it.

The guidance further defines what makes a firm's CHP effective. It should be:

- Fit for purpose – reflects the size and structure of the business;
- Available to all staff – intended to provide clarity and consistency to staff and clients;
- Understood by all staff – keep records of staff training;
- Readily shared with complainants or potential complainants – supplying a copy should be routine;
- Regularly reviewed at a senior level – record evidence of review, including reviewer details and review date;
- Agreed with PII brokers/provider(s) – should reflect processes that do not compromise PII cover;
- Include independent redress if the firm cannot resolve the complaint.

Concerning complaints handling, the RICS cannot decide whether a surveyor's opinion is correct or offer a second opinion. They also cannot resolve issues subject to court proceedings or where there is another resolution mechanism, e.g. challenging an Award under Party Wall etc. Act 1996.

RICS only considers complaints about service or professionalism, i.e., what action would a reasonable professional take? They will investigate matters if they are in the public interest and take disciplinary action to protect the public (rather than to punish the professional or firm). This process takes approximately six months, and there is no right to appeal the decision. A complainant can request a review by the Head of Regulation if a case is closed. However, the decision will be final.

How Can You Demonstrate Levels 1 and 2?

Your assessors will start at the highest level you have declared, which will be level 2.

At level 2, RICS expect that you will be able to provide practical experience in some of the following activities or tasks:

- Establishing a client's objectives;
- Defining a client's brief, including analysing desktop research or due diligence;

- Documenting a scope of services;
- Calculating fees;
- Compiling an appointment document or Terms of Engagement;
- Establishing who a project's stakeholders are and how you will manage them;
- Setting up a communication strategy with a client and applying this throughout the instruction;
- Issuing a client report;
- Dealing with a complaint;
- Measuring and reporting on KPIs, including improving performance where required;
- Providing a high standard of client care.

Typical questions will start with, 'How have you . . . ?', 'Tell me about an example when you . . .' or 'How did you provide good client care when . . . ?'. Remember, all the assessors' questions should be based upon the examples you provide in your level 2 write-up for this competency in your summary of experience – rather than being hypothetical or general.

You may also be asked some level 1 questions, focusing on the knowledge behind your practical level 2 examples. An example of this might be, 'tell me about the requirements of the RICS relating to complaints handling, Terms of Engagement or money laundering checks', for example.

Reference List

APM, 2023. *What Is Stakeholder Engagement?* [Online] Available at: www.apm.org.uk/resources/find-a-resource/stakeholder-engagement/ [Accessed 23 July 2023].

Chapter 4

Communication and Negotiation

Communication and Negotiation is a mandatory RICS APC competency required to level 2. This means that you need to write up your summary of experience, including a statement at both levels 1 and 2.

The RICS pathway guide sets out the relevant knowledge (level 1) and practical application (level 2) for this competency. However, you do not need to have experienced everything listed, as this will depend on your experience and role.

This chapter will explain what you should know in relation to Communication and Negotiation, covering the main requirements of level 1. Communication and negotiation will be dealt with separately, as they are two distinct activities – although every negotiation also involves communicating effectively.

Communication

Effective communication is a key skill that every surveyor needs to master, as it forms the basis of all professional instructions and interactions. Surveyors need to communicate with a wide variety of other people, ranging from lay clients to other highly experienced professionals. This means that surveyors need to be able to adapt their communication method and style to the audience in question.

Effective communication requires:

- A communicator;
- A method;
- A recipient.

DOI: 10.1201/9781003437116-4

However, misunderstandings and confusion can be created within all three elements:

- The communicator has to encode their message with the chosen communication method – does this reflect the original thought, intended message and purpose of the communication?
- Is the method appropriate for the intended message?
- The recipient has to decode the communication – does the understood message reflect the message that the communicator intended to relay to the recipient?

Communication can also be one way or two way. In one-way communication, we do not receive any immediate feedback, e.g. a letter. In two-way communication, we receive immediate feedback and can clarify or rectify misunderstandings, e.g. verbal communication. This, of course, depends on whether the other party is listening, though, which is a skill in itself. Deciding on whether one-way or two-way communication will be most effective is important – writing an email or a letter (one way) might be better if clarifying client instructions or a rent review proposal, whereas calling another agent to discuss a deal (two-way communication) may result in a better outcome. Difficult or challenging conversations or negotiations are often better delivered via two-way communication, as it allows both parties to raise concerns, dispel misunderstandings and build rapport.

Early career stage surveyors are often keen to 'hide' behind email, as they lack the confidence to speak to another more experienced surveyor or client. However, practice makes perfect, and it is advisable to work on your communication skills early in your career. Some of the relationships you build at this stage will remain with you for years to come.

With the advent of working from home post-pandemic, it is really important to ensure that you have the opportunity to learn from your peers. Being in an office or shadowing another surveyor provides an excellent opportunity to pick up skills from others. These could include overhearing how to deal with a difficult situation or how to negotiate on the telephone. An office environment also provides you with the opportunity to share ideas with others, discuss instructions, receive feedback on your work and pick up the key skills a surveyor needs. These are often not written in any university-level textbook.

When communicating, there are some key questions that surveyors should ask themselves:

- Who is the target audience?
- Are there any time limits or urgency of response required?
- Does the communication need to be recorded in writing? (for example, a rent review Calderbank offer);
- Is there a history of communication between the parties?
- What is the most appropriate method of communicating?
- What is the intended message?

Communication comes in a variety of forms, including:

- Written – reports, letters, emails, social media, books, magazines, online articles and blogs;
- Graphic – maps, drawn plans, sketch notes, designs, logos, branding and visualisations;
- Verbal – telephone calls, meetings, tenders, presentations, managing people and negotiations (more about this later on in this chapter);
- Non-verbal – body language, appearance, posture, eye contact, facial expressions and gestures.

These are some of the biggest barriers to effective communication; see if you can come up with ways to avoid these in your professional practice:

- Verbal – tone of voice, clarity, language barriers and what we don't say (i.e., silence);
- Jargon/technical language – who is the audience? Will they understand technical terms, or should simple and non-technical/lay language be used?
- Emotional/cultural barriers;
- The recipient is disinterested or not paying attention;
- Timing of the communication;
- Physical barriers, e.g. arms folded or hand over mouth;
- Differences in perception/viewpoint;
- Prejudice/bias;
- Differing expectations;

- Interruptions – noise, physical distractions;
- Location – cannot physically meet in person;
- Attitude/mood;
- Poor listening skills;
- Assumptions/prior experience;
- Ambiguity;
- Context.

Active listening is essential to facilitate effective communication. This goes beyond just hearing what is being said. Active listening involves being present and active in the communication process. Skills include using eye contact, employing non-verbal cues, asking open-ended questions, paraphrasing, reflecting and being attentive.

Where communication is ineffective, this can lead to time delays, frustration, stress, unforeseen costs and contractual issues. Thus, getting it right can lead to a more positive experience, potentially winning new work, doing a difficult deal or resolving a complex situation.

Here are ten tips on ways to be a better communicator:

1. Listen – in a conversation, try to listen more than you speak. You can learn a lot about a person or situation this way – encourage people to talk to you and acknowledge what they say;
2. Use silence – often, this can communicate a message better than words. Leave a gap in conversation – what the other person says to fill this can often tell you a lot;
3. Check your body language – try to maintain a relaxed, open stance and friendly demeanour. Appearing defensive or closed can prevent effective communication or deter others from engaging in productive dialogue;
4. Clarity – be clear and concise in what you say or write, avoid using unnecessary words and get to the point quickly where possible;
5. Timing – don't ask an important question first thing on a Monday morning or at 5 minutes to close on a Friday afternoon. Ask when your recipient has time to consider your communication – if they don't, ask when would be a good time to communicate;
6. Don't hide behind emails – use the phone and speak to your negotiating counterparts or clients rather than sending a bland email. You are often able to explore alternative opportunities or other ways to overcome challenges. It is often also a lot quicker to reach a resolution or agree on a course of action;

7. Feedback – adapt your communication to the feedback you receive. Maybe it would be better to meet your recipient in person or to follow up on your conversation with an email. Or, maybe you need to clarify what you said to prevent a misunderstanding;

8. Ask open-ended questions to learn more about your recipient and enable you to communicate more effectively;

9. Remember to say the other person's name – it is one of the most emotionally powerful words to each of us. You don't need to say it often, but you do need to consider how you say it – we suggest sincerely and positively;

10. Learn from your mistakes – if a communication does not go well, ask for feedback and analyse what went wrong. How could you do it better next time? You could ask trusted colleagues, family or friends for feedback on your communication style and impact.

Another key communication skill is emotional intelligence. This is a surveyor's ability to monitor their emotions, to differentiate emotions and to label them appropriately. Emotional intelligence can be used to guide thinking, behaviour and communication. This is an area that you may be interested in undertaking further training.

In relation to your RICS APC submission, make sure you proofread your written work carefully. Asking a non-surveyor is a great way to do this. They will often identify issues that the surveyor does not, and they can also check if the overall 'story' of examples and submission make sense.

Another great tip is to use the Spelling & Grammar check function on Word or a more comprehensive tool such as Grammarly.

Here are some top tips for ensuring you use appropriate grammar in your work:

1. Long sentences – it can be really difficult to read a submission where extremely long sentences are used. Try to keep sentences short and to the point, breaking them up at a natural point. For example, the two preceding sentences could have been written as 'Long sentences – it can be really difficult to read a submission where extremely long sentences are used, so try to keep sentences short and to the point and break them up at a natural point'. This feels a little clumsy and is harder to follow;

2. Use of contractions – the RICS APC submission is a formal piece of work. We recommend avoiding contractions such as 'don't'

or 'can't' – instead, use the full 'do not' or 'cannot'. It might be tempting to use these to save words, but it reduces the overall professionalism and formality of your written work;

3. Use of US rather than UK English (if you are a UK candidate) – always use UK English if you are submitting in the UK, so use 's' instead of 'z'. For example, 'formalise' instead of 'formalize'. There are also some specific spellings to be aware of, e.g. 'meter' in the US and 'metre' in the UK. Always check that the spell check function on your word processing software is set to English (UK) to help avoid these errors;

4. Use of apostrophes – we often see apostrophes for singular and plural words being used incorrectly. If you are talking about something that a client owns, then you need to use the singular 'the client's property'. If you are talking more generally about meeting the needs of multiple clients, then you would use 'clients' needs';

5. Tense – in the RICS APC submission, ensure you use the first person and the past tense. This means that you will be talking about what you did – not what someone else is doing or what your firm might do! For example, 'I dealt with a rent review in Bristol. I inspected, measured and analysed the subject lease', rather than 'We dealt with a rent review in Bristol. We would inspect, measure and analyse the subject lease'. You need to clearly explain in your submission what you did (not what others did) in real-life examples (which will have happened in the past);

6. Being too verbose – a key skill in writing your RICS APC submission is to be clear and concise. Sometimes we see phrases that are too verbose and could be simplified. For example, 'It came to my attention that the client's rent review was overdue, and I advised them to instruct me to deal with it at the earliest opportunity'. Instead, it would be much better to say, 'I identified an overdue rent review and sought instructions from my client to deal with it'. This is where asking a non-surveyor to proofread your work will pay dividends;

7. Use of 'and' in a list – if you are writing a list, then you only need 'and' before the last item in the list. So, for example, the correct way would be to write 'I collated, weighted and analysed the comparable evidence', rather than 'I collated and weighted and analysed the comparable evidence'. Whilst this is a simple example, it is important to note the correct use of 'and' in longer lists;

8. Use of 'could of' rather than 'could have' – although when spoken, 'could have' sounds a lot like 'could of' – the correct usage is always 'could have'. So, for example, 'The client could have proceeded with option one. However, the time constraints meant that I advised proceeding with option two';
9. Tautologies – this is where you say the same thing twice in a sentence, just using different words. For example, 'dilapidated ruins', 'close proximity', 'added bonus' and 'large crowd'. These all say the same thing twice – you could just say 'ruins', 'close', 'bonus' and 'crowd';
10. You and you're – getting the use of 'your' and 'you're' and 'their' and 'there' can be tricky. 'Your' is where something belongs to you. 'You're' means 'you are', so saying something like, 'You are an APC candidate'. 'Their' is similar in that this is where something belongs to another person (or them). 'There' relates to where something is physically, so 'over there'.

Excellent additional resources on written communication can be found on the Plain English Campaign website. They have guidance specifically available on report writing, letters and emails.

In relation to reports, they should be appropriate for the client's requirements. These should be agreed upon at the outset, e.g. short or long form and presentation requirements. Using spreadsheets, graphs, charts and images can be a good way to bring a report to life and communicate more effectively with the client or end user. Ensure that all figures or tables are titled clearly in the text and on the contents page. There are many excellent sources of information and guidance online on how to use Microsoft Excel (or other spreadsheet packages) effectively, including Philip Bowcock's book, *Excel for Surveyors*.

In relation to letter writing, a key challenge is often the use of 'Yours faithfully' (where you do not state the recipient's name) or 'Yours sincerely' (where you state the recipient's name). Formal language appropriate to the audience should be used, considering whether the use of lay or technical terms is more appropriate.

Surveyors are highly likely to be client-facing, either through informal interactions (phone calls, email updates or site meetings) or formal meetings or pitches.

Client updates should be issued in a professional, consistent and readable format, such as a spreadsheet or word-processed report.

Reporting should be frequent, and any key issues or risks should be highlighted at the earliest opportunity.

In formal meetings or pitches, surveyors may need to give a presentation or speech to clients. Giving a presentation can be nerve-wracking, and finding ways to calm nerves and boost confidence is essential. Surveyors should rehearse what they are going to say and perhaps take brief notes or cue cards to jog their memories during the presentation. Positive body language and clear verbal delivery are also key to communicating clearly.

Ways to calm your nerves before a presentation include:

- Use eye contact to connect with your audience;
- Practice in front of a mirror, record yourself or ask friends, family or colleagues for feedback;
- Practice using breathing techniques;
- Visualise the experience beforehand;
- Avoid too much caffeine beforehand. It can make you feel more nervous or jittery;
- Have a glass of water with you;
- Smile.

The use of a good visual aid may help the surveyor to communicate complex ideas or concepts effectively using presentation software such as Microsoft PowerPoint.

Here are some of our tips for creating an effective presentation:

- Keep text to a minimum. The five/five/five rule could be followed, where no more than five words per line of text, five lines of text per slide or five text-heavy slides are used in a row;
- Graphics, charts and other visuals are a great way to retain attention and make slides visually interesting;
- Keep presentations clear and concise. The 10/20/30 rule could be followed, where a presentation has no more than ten slides, lasts no more than 20 minutes, and only contains fonts larger than size 30;
- Only discuss one idea on each slide;
- Keep slide presentation consistent using a master slide and templates for various types of slides. Ensure fonts and font sizes remain consistent throughout;
- Use a colour scheme that is accessible, e.g. no red on green/green on red (which is not accessible for those who are colour blind) and dark-coloured text on a light background.

In your RICS APC final assessment interview, you will need to give a presentation on your case study (career history if you are a senior professional candidate) in the first ten minutes. During this, you will be assessed on your communication skills, and this will go some way towards ensuring that you hit level 2 in this competency. You can find a downloadable template case study presentation template at www.property-elite.co.uk/free-resources (Property Elite, 2023).

The current online format of the APC interview requires a slightly different approach to a presentation in person, as you are one step removed from your assessment panel. Practising giving your presentation online beforehand can be a good way to ensure you are fully prepared for the day itself. Avoid hiding behind your visual aid and ensure that you are able to present without it in the unfortunate case of any IT issues.

Negotiation

We will now move on to the other half of this competency: negotiation. Many candidates forget to include negotiation within their level 1 and 2 write-up for this competency, which can be a referral point if you are on the preliminary review route.

Negotiation is defined by the Cambridge Dictionary as 'The process of discussing something with someone in order to reach an agreement with them, or the discussions themselves' (Cambridge Dictionary, 2023).

The role of a negotiator is defined by RICS in the Practice Statement and Guidance Note Surveyors Acting as Expert Witnesses 4th Edition (RICS, 2023) as

Acting to negotiate a resolution to disputed matters. In such a role, you will have no involvement with a tribunal except insofar as you or others may perceive a possibility that a failed negotiation may then necessitate a reference to a tribunal, at which point you or another professional may be engaged to act as an advocate or provide expert evidence as an expert witness. It is possible that some negotiators may not find it possible to act as an expert witness as their impartiality may be damaged, or may be perceived to be damaged, by their prior or continuing role of negotiator. It is recommended that you be alert to this.

The process of negotiation can be simplified to:

- Preparation – it is essential not to forget this stage. Generally, a better-prepared negotiator secures a better outcome;
- Discussion or information exchange – this is where the negotiation begins, and the parties can explore each other's positions, interests and objectives;
- Bargaining – this is where the negotiation really happens, and effective communication is key;
- Conclude and agree – this is where the negotiation outcome is formalised, ideally in writing;
- Implement and execute – this is where the parties follow through on their agreement. Doing so helps to maintain a good relationship and can facilitate future negotiations.

Candidates also need to have an understanding of basic negotiation theory.

This includes various negotiating styles, such as:

- Accommodating – seeking to maintain the parties' relationship and minimising conflict during negotiation;
- Avoiding – remaining objective and often withdrawing from the negotiation (not recommended);
- Collaborating – trying to problem solve with the other party and achieve an outcome where both parties feel they have 'won';
- Competing – focusing only on your interests and being unwilling to compromise. This typically starts with making a low offer and then increasing offers until the matter is settled;
- Compromising – seeking a middle ground that works for both parties rather than an outcome where both parties feel they have 'won'.

Many RICS-accredited degree courses teach the concept of principled negotiation, popularised by the book *Getting to Yes* by Fisher & Ury. Principled negotiation takes a collaborative approach and focuses on mutual gains. There are four key elements of this theory including:

- Separating people from the problem and removing the 'emotion' from the negotiation;
- Focusing on the parties' underlying interests (i.e., needs, wants and motivations) rather than positions;

- Finding options for mutual gain, which often requires a little creativity and deeper exploration into issues;
- Basing the negotiation on objective criteria or evidence.

The concepts of BATNA (best alternative to a negotiated agreement) and WATNA (worst alternative to a negotiated agreement) are useful negotiation tools. BATNA is essentially the fallback position if an agreement cannot be reached through negotiation, such as going to court or withdrawing from a transaction completely. WANTNA is the worst-case scenario and can help to focus a party on reaching a negotiated outcome. An example could be losing the opportunity to do business with the other party.

Key skills for surveyors to develop when negotiating, irrespective of approach or strategy, include:

- Discussing and agreeing the client's objectives beforehand, including negotiable and non-negotiable points;
- Carrying out detailed research, preparation and a SWOT analysis;
- Preparing 'win-win' and fallback positions;
- Discussing areas of common ground during negotiations;
- Creating a constructive environment to negotiate within;
- Deciding on the best method of communication during negotiations, e.g. a meeting followed up by an email.

How Can You Demonstrate Levels 1 and 2?

Your assessors will start at the highest level you have declared, which will be level 2.

At level 2, RICS expect that you will be able to provide practical experience in some of the following activities or tasks:

- Writing letters, reports and formal documents;
- Taking and recording minutes of meetings;
- Producing pricing documents;
- Delivering reports at meetings;
- Taking part in interviews;
- Giving presentations to staff, project teams or clients;
- Negotiating a transaction, settlement, loss and expense claim, extension of time, acceleration programme, contract sum or final account;
- Agreeing the value of an instruction.

Typical questions will start with, 'How have you . . . ?', 'Tell me about an example when you communicated or negotiated effectively . . .' or 'How did you communicate or negotiate well when . . . ?'. Remember, all of the assessors' questions should be based upon the examples that you provide in your level 2 write-up for this competency in your summary of experience – rather than being hypothetical or general.

At level 2, in your summary of experience, your assessors will expect to see a balance of examples. Ideally, this will be one that focuses on communication and a second that focuses on negotiation. Candidates sometimes struggle with level 2 examples of negotiation. However, all candidates are likely to have some relevant experience, such as negotiating a fee or contract with a client, negotiating a transaction price or agreeing a settlement.

You may also be asked some level 1 questions, focusing on the knowledge behind your practical level 2 examples. Be prepared to explain the theory or concepts behind how you communicated or negotiated effectively. Also, remember that your communication skills will be assessed within your written submission, case study presentation and the rest of your final assessment interview. This, in itself, could be sufficient to allow you to demonstrate competence to level 2. Don't neglect the quality of your submission, presentation or interview responses – prior preparation, practice and proofreading will help you to achieve the required standard.

Reference List

Cambridge Dictionary, 2023. *Negotiation*. [Online] Available at: https://dictionary.cambridge.org/dictionary/english/negotiation# [Accessed 27 July 2023].

Property Elite, 2023. *Free Resources*. [Online] Available at: www.property-elite.co.uk/free-resources [Accessed 27 July 2023].

RICS, 2023. *Practice Statement Surveyors Acting as Expert Witnesses*. [Online] Available at: www.rics.org/content/dam/ricsglobal/documents/standards/Surveyors%20acting%20as%20expert%20witnesses_Feb2023amend.pdf [Accessed 27 July 2023].

Chapter 5

Health and Safety

Health and Safety is a mandatory RICS APC competency required to level 2. This means that you must write up your summary of experience including a statement at levels 1 and 2. Some pathways allow candidates to take Health and Safety as a technical competency to a higher level, i.e., level 3.

You need to know about relevant RICS guidance and key legislation, including how these affect your actions on-site and in the workplace. You also need a strong understanding of how to work safely and to take responsibility for the health and safety of those that you have a duty of care for. There may be health and safety considerations specific to your area of practice, such as fire safety, working at height or asbestos risk.

The RICS pathway guide sets out the relevant knowledge (level 1) and practical application (level 2) for this competency. However, you do not need to have experienced everything listed, as this will depend on your experience and role.

This chapter will explain what you should know concerning Health and Safety, covering the main requirements of level 1.

Legislation

The key primary legislation relating to health and safety at work in the UK is the Health and Safety at Work etc. Act 1974. We will also discuss a wealth of secondary legislation or statutory instruments which provide further sector-specific legislation.

The 1974 Act is enforced by the Health & Safety Executive (HSE), alongside local authorities. It requires employers to, as far as is

DOI: 10.1201/9781003437116-5

reasonably practicable, ensure employees' health, safety and welfare at work. This includes staff training, welfare provision, safe working environments and the provision of information, which is reflected in the RICS guidance that we will discuss later in this chapter. If an employer has more than five employees, they must have a written health and safety policy.

An employer's breach of the 1974 Act constitutes a criminal offence and can lead to an unlimited fine and/or up to two years' imprisonment. A director or manager can also be liable for their company's breach if it was committed with consent or through neglect. An employee can also be held personally liable if they intentionally or recklessly interfere with or misuse anything provided concerning health and safety.

The Workplace (Health, Safety and Welfare) Regulations 1992 apply to most workplaces (apart from construction sites or involving construction work) and cover various health and safety issues, such as ventilation, workstations, falling objects and windows.

The Corporate Manslaughter and Homicide Act 2007 was a seminal piece of legislation defining the criminal offence of corporate manslaughter (or corporate homicide in Scotland). This is where the actions of a company (e.g., management failure) lead to a gross breach of their duty of care and the death of an employee or contractor. The penalties include an unlimited fine, imprisonment and disqualification of a member or senior management team members as a company director or directors.

Surveyors must also be aware of the Occupiers Liability Act 1957 and 1984.

The 1957 Act confirms that the occupier or a landlord in control of premises must take reasonable care to ensure the reasonable safety of visitors (who have implied or express permission to be there or a legal right of entry). The level of care depends on the nature of the visitors, so a much higher duty is owed to children than skilled visitors (who should be aware of the risks and measures to prevent harm). Under the 1957 Act, adequate warnings and signage may discharge the liability if they enable visitors to remain reasonably safe on-site.

The 1984 Act applies to trespassers rather than visitors. Trespassers are not invited onto a property, and their presence is unknown and not consented to. The 1984 Act applies where the occupier or a landlord in control of premises is aware of a danger and has reasonable grounds to believe that a trespasser may come into the vicinity of this danger. A common duty of care is established to take reasonable care for trespassers, although this is a lower duty of care than is owed to visitors under the 1957 Act.

While the 1957 Act permits claims for death, personal injury and property damage, the 1984 Act only permits claims for the first two (i.e., not for property damage).

Another essential health and safety issue that surveyors must be aware of is asbestos. Simply put, asbestos kills, so we must be mindful of the risks to advise our clients diligently and keep ourselves and others safe on-site.

Asbestos is a naturally occurring hazardous material (a fibrous mineral silicate) and harmful to health. If disturbed, it releases fibres into the atmosphere, which, if inhaled, can lead to asbestosis, a form of cancer.

Asbestos comes in three forms: crocidolite (blue), amosite (brown) and chrysotile (white). Blue and brown asbestos were banned in the UK in 1985, whilst white asbestos was finally banned in 1999.

Asbestos was historically used in many building materials. If a building was constructed pre-2000, you should always assume that asbestos may be present.

Common asbestos-containing materials (ACMs) include:

- Acoustic plaster;
- Adhesives;
- Ceiling tiles;
- Cement pipes;
- Decorative plaster;
- Ducting;
- Fire curtains and doors;
- Fireproofing;
- Pipe insulation;
- Roof felt;
- Textured paint (such as Artex) and coatings;
- Vinyl floor tiles;
- Wallboard.

Regulation 4 of the Control of Asbestos Regulations 2012 imposes a legal duty to manage asbestos if it is present or presumed to be present. The duty holder under the Regulations will either be the owner (e.g., of vacant premises) or occupier (e.g., a tenant with an FRI lease), responsible for repairing and maintaining commercial premises and the common parts of domestic premises. In multi-let properties, the duty may be shared between both parties.

Where a managing agent is instructed, the landlord (duty holder) cannot delegate their statutory duty to manage asbestos. However, the landlord can claim (contractually or in tort for a non-contractual duty) against the managing agent if any action or inaction contributes to the landlord's failure to comply as a duty holder. The managing agent will also have a duty to manage asbestos risk under the Health & Safety at Work etc. Act 1974.

If the landlord is absent, the duty holder may acquire additional duty holder responsibilities as they may be considered to be in charge of the property.

Where the duty to manage under Regulation 4 does not apply, a duty arises (under Sections 2 and 3 of the Health & Safety at Work etc. Act 1974) to minimise the risk of asbestos in domestic premises. The Defective Premises Act 1972 and the Homes (Fitness for Human Habitation) Act 2018 also emphasise this duty in domestic premises.

In the event of a substantial breach of the Regulations, the penalties include an unlimited fine and up to two years' imprisonment.

If a property is suspected to contain asbestos, it is a legal requirement to commission an asbestos survey and report. This will identify and record the location, amount and type of any ACMs.

A management plan should then be prepared to state how the risk of any ACMs present will be managed, including an asbestos register recording the location and condition of the ACMs. This will be used to manage ACMs during everyday occupation and use. Appropriate action should be taken to manage, repair or remove the asbestos in line with the survey and report.

The asbestos register should be updated annually or more frequently if the risk of deterioration of the ACMs is high. The management plan should be reviewed every six months or after refurbishment, change of use or changes in company procedures.

The management plan and register should be provided to anyone inspecting or working on the property. However, if the premises are to be refurbished or demolished, then a refurbishment/demolition survey will be required.

If asbestos needs to be removed, it will be categorised as licensable (higher risk, e.g., work involving loose fill insulation or asbestos millboard) or non-licensable (lower risk, e.g., maintenance work involving asbestos cement products or asbestos-containing vinyl floor tiles). A complete list is provided on the HSE website.

For licensable asbestos, an HSE-licensed contractor must be used and the work notified using an ASB5 form, at least 14 days before works start.

To remove non-licensable asbestos, contractors do not have to be licensed. However, they must use suitably trained workers to carry out the work.

Some non-licensed work must, however, still be notified to the HSE, such as minor, short-duration work involving asbestos insulating board or insulation. Contractors must be placed under health surveillance by their GP, and written records kept of the work.

ACMs must be disposed of at an Environment Agency licensed asbestos site in all cases.

Surveyors must be familiar with the RICS Guidance Note Asbestos: Legal Requirements and Best Practice for Property Professionals and Clients (4th Edition). This confirms that all surveyors should have undertaken asbestos awareness training, as per Regulation 10 of the Control of Asbestos Regulations 2012.

Other important health and safety issues to be aware of, particularly concerning property management, include:

- Lifting Operations and Lifting Equipment Regulations 1998 – this requires lifting equipment to be kept in a safe condition, safely positioned and installed and used safely;
- Smoke-free (Premises & Enforcement) Regulations 2007 – this banned smoking in substantially enclosed workplaces and public places. The Regulations also require no smoking signage to be displayed in these places;
- Legionnaires' disease – this is a potentially fatal form of pneumonia caused by legionella bacteria, released into the air by purpose-built water systems, e.g., air conditioning systems, evaporative condensers and hot tubs. Employers or those in control of premises must ensure that a specific risk assessment is carried out and systems are maintained correctly. This could include keeping water temperatures between 20°C and 45°C and regularly servicing systems.

The Construction (Design & Management) Regulations 2015 (CDM) are a key piece of legislation for all surveyors to be aware of. They aim to improve health and safety standards, management and coordination in the construction industry. CDM applies to all construction work on non-domestic and domestic properties.

There are various roles defined under CDM (Figure 5.1).

Under CDM, work is notifiable to the HSE via an F10 if it:

- Lasts over 30 days and has over 20 workers simultaneously working at any point; or
- Exceeds over 500 person days of construction work.

If a project is not notifiable, then the requirements of CDM still apply in full.

Section 11 of the Landlord and Tenant Act 1985 sets repair standards for residential property leased for less than seven years. This relates to the structure, exterior, mains services supplies, sanitation, heating and heating water.

The Housing Health and Safety Rating System (HHSRS) was introduced by the Housing Health and Safety Rating System (England) Regulations 2005 to meet the requirements for assessing hazards under the Housing Act 2004.

The HHSRS is a risk-based evaluation tool that local authorities can use to identify and manage health and safety risks in rented residential property, e.g., damp, mould, excessive cold, overcrowding, slip and trip hazards, faulty gas appliances, fire risks and dangerous electrical installations.

The most serious hazards, e.g., leaking roofs, rats or broken steps, are allocated a Category 1 rating and pose a serious threat to the health and safety of occupants and visitors. The local authority must take enforcement action concerning these hazards, such as hazard awareness notices, improvement notices, emergency remedial actions, prohibition orders and demolition orders.

The Gas Safety (Installation & Use) Regulations 1998 require landlords to keep gas pipework, appliances, chimneys and flues in a safe condition. A Gas Safe registered engineer must undertake annual gas safety checks. The tenant must be given a copy before moving in, and subsequent checks must be provided to existing tenants within 28 days.

The Electrical Safety Standards in the Private Rented Sector (England) Regulations 2020 apply to residential lettings, including Assured Shorthold Tenancies (ASTs) and Houses in Multiple Occupation (HMOs).

Role	Duties
Client	• Appointing other duty holders and ensuring they carry out their duties; • Allocating time and resources; • Providing information to other duty holders, i.e., the Pre-Construction Information (PCI); • Providing welfare facilities. If a Client fails to appoint the other roles in writing, they are deemed to have appointed themselves (and will have to satisfy the duties of the other CDM roles).
Domestic client	Duties are generally transferred to the Contractor (single contractor project), or Principal Contractor (multi-contractor project). Otherwise, the Principal Designer can be appointed if agreed in writing.
Principal Contractor (multi-contractor projects)	• Construction phase planning, management, monitoring and coordination of health and safety; • Liaising with other duty holders; • Preparing the Construction Phase Plan (CPP); • Providing site inductions, preventing unauthorised access, ensuring work is carried out safely and providing welfare facilities.
Contractor	• Safely carrying out construction work; • Coordinating work on multi-contractor projects; • Following instructions from the Principal Contractor and/or Principal Designer; • Preparing a CPP on single contractor projects.
Principal Designer	• Pre-construction phase planning, management, monitoring and coordination of health and safety; • Identifying, eliminating or controlling foreseeable risks; • Ensuring Designers carry out their duties; • Providing information to the Principal Contractor for the construction phase; • Developing a Health & Safety File to hand over to the Client upon completion.
Designer (multi-contactor projects)	• Identifying, eliminating or controlling foreseeable risks when preparing or modifying designs; • Providing information to other duty holders.

Figure 5.1 CDM Roles

Electrical installations (including fixed wiring, but not portable appliance testing) must be inspected, and an Electrical Installation Condition Report (EICR) must be provided by a competent person (e.g., an electrician registered with NICEIC and/or ELECSA) every five years. Landlords must give this to new tenants before the start of their tenancy and to existing tenants within 28 days of the inspection.

The penalty for non-compliance is a maximum fine of £30,000. Local authorities can also serve enforcement notices, carry out required work and recover the cost from the landlord.

RICS Guidance

All surveyors must be familiar with the RICS Guidance Note, Surveying safely: health and safety principles for property professionals (2nd Edition). This provides best practice and highlights the corporate and personal responsibility to ensure that high standards of health and safety are maintained in the workplace.

A key concept introduced by the Guidance Note is that of the 'safe person'. This means that a surveyor must assume responsibility for their own safety, the safety of colleagues and the health and safety of others in the workplace.

The Guidance Note confirms that RICS Regulated Firms must provide a safe working environment, safe work equipment, safe systems of work and training to ensure that staff are competent. Staff must use these safe systems and equipment to protect themselves and others on-site.

This could include being provided with and wearing adequate and well-fitting personal protective equipment (PPE), under the Personal Protective Equipment at Work (Amendment) Regulations 2022. The provision of PPE applies to both employees and casual workers, known as limb (a) and (b) workers, respectively. This includes steel-toed and soled boots, hi-vis jackets, safety helmets, hard hats and gloves. The specific PPE required will depend on the work environment.

Regulated Firms must carry out risk assessments in relation to the office, workplace and sites that surveyors may visit. Surveyors must also carry out risk assessments before visiting or inspecting a site or property.

On-site, a dynamic risk assessment should be continually undertaken as the situation may be different or change on the day, and the surveyor's actions or activities may be modified as a result. An example of this could be a dog on-site, which the surveyor needs to ask to be contained in a separate room, or an unsafe loft ladder or limited access, which the surveyor cannot use or access safely.

Risk assessment involves assessing what could cause harm so that sufficient precautions can be taken to prevent harm. A hazard is defined as something which may cause harm. A risk is the likelihood of the harm happening.

Risk assessment usually involves the following steps:

* Identify the hazards;
* Identify who may be harmed and how they may be harmed;
* Evaluate the risks and decide on precautions to be taken;
* Record the findings and implement them;
* Review and update the risk assessment regularly;
* Advise those affected by the outcome of the assessment and methods of work, or other control measures necessary, to minimise or eliminate risk.

Candidates should also be familiar with the hierarchy of risk control. This involves the following:

* Eliminating – redesigning the activity or substituting a substance so that the hazard is removed, e.g. using a drone to avoid working at height;
* Substituting – replacing the materials used or the proposed work process with a less hazardous one, e.g. using pre-prepared components rather than cutting on-site;
* Engineering controls – using equipment or engineering controls to manage risk, e.g. using work equipment or other measures to prevent falls or separating the hazard from operators by enclosing equipment;
* Administrative controls – identifying and implementing procedures to work safely, e.g. reducing the need for lone working or ensuring that work is undertaken in daylight;
* Using PPE – only acceptable if the aforementioned measures cannot be used, e.g. emergency alarms could be provided where lone working cannot be avoided.

Firms will also need to carry adequate insurance, such as employer's liability (covering claims by employees who are injured or become ill at work) and public liability (covering claims by the public for personal injury or property damage) insurance.

Regulated Firms must lead on health and safety compliance from the top with clear written policies and procedures. All candidates should ensure that they read these and can discuss them at their interview.

Specific issues which might be raised in company policy include:

- Providing a safe place of work relating to ventilation, heating, lighting and welfare facilities;
- Minimising risks relating to monitors and workstations;
- Personal Protective Equipment;
- Manual handling;
- First aid;
- Reporting of Injuries, Diseases and Dangerous Occurrences Regulations (RIDDOR) 1995;
- Safe electrical systems;
- Control of Substances Hazardous to Health Regulations 2002;
- Asbestos;
- Fire risk assessment;
- Working hours;
- H&S induction;
- Driving;
- Stress;
- Lone working.

Lone working is an issue which is essential to manage. In 1986, Suzy Lamplugh, an estate agent, went missing after going to meet an unknown client. Firms and surveyors need to be aware of the risks of lone working – if it can be avoided, then it should be, in line with the hierarchy of risk control. If it cannot, then these are some precautions that surveyors can take:

- Taking a charged mobile and personal alarm;
- Planning an escape route;
- Implementing a call-back system with the office (e.g. a safe word);
- Making a daily schedule available to colleagues;
- Being careful in roof voids and when using ladders;
- Parking your car close by and keeping your keys on you;
- Making sure you know who you are meeting;
- Following your gut instinct;

- Understanding the site rules for construction sites;
- Being aware of aggressive occupants and dogs;
- Informing others of any medical conditions.

If in doubt, however, always withdraw from the situation or inspection. You can always return later with a colleague or when the problem has been resolved, or you can record a limitation in your report or client advice. If you are asked to inspect a site alone, and you are concerned, take a stand and refuse – report this internally and only inspect if you feel it is safe. Your safety and life are at risk – and this is far more important than any piece of work, project or instruction will ever be.

Working at height is one of the leading causes of injuries and fatalities, including falling from ladders and walking on fragile roof coverings. The key legislation in this respect is the Work at Height Regulations 2005. Risk assessment is required before any inspection involving working at height. This includes considering the height required, duration of the inspection, frequency of inspection and the roof or surface condition.

Applying the hierarchy of risk control, working at height should be avoided where possible, e.g. using a telescopic pole and camera or a drone, instead. If working at height cannot be prevented, then action to prevent falls should be taken and the risk of a fall minimised. This could involve installing a guardrail, anchor points (attached to a safety harness), using a fall arrest system, scissor lift, ladder or tower scaffold. Scaffolding should ideally be tagged and dated after inspection, confirming it is safe.

Ladder safety is something that all surveyors should be trained in. The Home Survey Standard, which applies to residential level 1, 2 and 3 surveys, confirms that a ladder (at least 4m long) can be used to inspect flat roofs and loft spaces where the access or hatch is no more than 3m above ground or floor level. The top section of the ladder should extend 1m above where the surveyor steps off the ladder. The condition of a ladder should be checked before each use. If the ladder is telescopic, care should be taken to ensure that each extendible section is locked into place securely before use. The HSE website provides excellent additional guidance on ladders.

Finally, the RICS published a Guidance Note on Health & Safety for Residential Property Managers (1st Edition). This covers a wide variety of compliance areas for residential property, which tend to be more onerous than commercial premises due to the more intensive use (including occupiers sleeping and living in the property).

A managing agent's obligations for a residential block will depend on the type, age and size of the block, the tenancy agreements in place and the individual nature of each tenant or leaseholder.

For long leasehold apartments, a service charge is typically paid by the leaseholders concerning the management and repair of the block's common areas. The freeholder (who may pass this liability to their managing agent) is responsible for the common areas using the service charge funds. The individual leases will define the service charge regarding common areas, services and apportionment. The freeholder or managing agent is generally responsible for health and safety in the common areas but not the internal parts of the apartments or flats (as they are in the control of the leaseholder).

The freeholder could let the apartments or flats directly on assured shorthold tenancies in a multi-let block. The maintenance liability remains with the freeholder rather than being passed onto the tenants. In this case, no service charge will be operated for the block, and the responsibility for the health and safety of the whole block will fall to the freeholder or managing agent.

Candidates in the construction industry may need accreditation under the Construction Skills Certificate Scheme (CSCS). Other schemes to be aware of are the Contractors Health & Safety Assessment Scheme (CHAS), Safety Schemes in Procurement (SSIP), Safe-Contractor and the Considerate Constructors Scheme (CCS).

Fire Safety

Fire safety is highly topical following the Grenfell Tower tragedy and extensive subsequent legislation. Fire safety is also a separate technical competency, which will be covered in other pathway-specific books. However, this chapter will provide a general overview of the critical things surveyors need to know.

Fire needs three things to start:

- An ignition source, e.g., heaters, lighting, naked flame or electrical equipment;
- A fuel source, e.g., wood, waste packaging or furniture;
- An oxygen source, i.e., the air.

The Regulatory Reform (Fire Safety) Order 2005 covers general fire safety relating to non-domestic property in England and Wales, including the common areas of multi-occupied buildings.

Non-compliance penalties are up to £5,000 for minor breaches, whereas major breaches carry unlimited fines and up to two years' imprisonment.

The Fire Safety Act 2021 was enacted pursuant to the Fire Safety Bill following the Grenfell Tower fire in 2017. It amends the Regulatory Reform (Fire Safety) Order 2005 (Fire Safety Order) to cover specific elements of buildings containing more than two separate residential units (irrespective of height or number of storeys). These elements are the structure, external walls (cladding, balconies and windows) and all doors (between flats and common parts).

The local authority enforces the 2005 Order. They can visit the premises and issue informal or formal fire safety alterations, enforcement or prohibition notices if measures are found to be insufficient.

The 2005 Order emphasises risk assessment and fire prevention, with a requirement to take reasonable steps to reduce the fire risk and ensure that occupants can escape safely if a fire occurs.

The 2005 Order allocates responsibility for fire safety to the Responsible Person, who is generally the employer or controller of the premises. The latter could be the owner, landlord, occupier or anyone else controlling the premises, such as a facilities manager, building manager, managing agent or risk assessor.

On a construction site, the Responsible Person is likely to be the main or Principal Contractor in control of the site. If there is more than one Responsible Person, they must liaise to ensure compliance with the Order.

The duties of the Responsible Person include:

- Carrying out and regularly reviewing a Fire Risk Assessment (FRA) (including updating it if required to reflect the provisions of the Fire Safety Act 2021). A written copy must be made available if a business has more than five employees;
- Communicating fire safety risks to employees and users of the premises;
- Installing and maintaining fire safety measures, including fire detection, warning systems and fire fighting equipment;
- Carrying out emergency escape planning;
- Providing staff training, including fire marshals and fire drills.

The process of preparing an FRA includes the following:

- Identifying fire hazards;
- Identifying people at risk;
- Evaluating, removing or reducing risks;
- Recording findings, preparing an emergency plan and providing training;
- Reviewing and updating the FRA.

Where there is a high risk of fire relating to the external walls or cladding of a multi-occupied residential building, then a further fire risk appraisal must be undertaken in line with PAS 9980: Fire risk appraisal and assessment of external wall construction and cladding of existing blocks of flats – Code of practice.

The Fire Safety (England) Regulations 2022 place additional duties on the Responsible Person or Persons for mid- and high-rise blocks of flats to:

- Provide information to the fire and rescue services, including electronic building plans and construction detailing for the external walls and any cladding;
- Undertake monthly checks of fire fighting or fire escape equipment;
- Install a secure information box and escape route signage;
- Provide residents with fire safety information.

The Building Safety Act 2022 has also been enacted and substantially amends the 2005 Order. Three new regulatory bodies have been introduced; the Building Safety Regulator, the National Regulator of Construction Products and the New Homes Ombudsman.

A new duty holder, the Accountable Person, has been introduced for buildings over 18m in height or with over seven storeys. The Accountable Person is responsible for fire and structural safety within the building.

Residential valuers must also know about External Wall System (EWS) 1 forms and the current RICS guidance. This will be covered at length in the Residential pathway book in this series.

Other key legislation in the residential sector includes the:

- Furniture and Furnishings (Fire Safety) Regulations 1988 (as amended in 1989, 1993 and 2010) – these require any furniture or furnishings, except for mattresses, bed-bases, pillows and cushions, to have appropriate fire safety labels;

- Smoke and Carbon Monoxide Alarm (England) Regulations 2015 (as amended in 2022) – this requires landlords to install and maintain a smoke alarm on each floor of residential dwellings and to install carbon monoxide alarms in any living areas containing a fixed combustion appliance (excluding gas cookers).

Surveyors may also need to be aware of the International Fire Safety Standards (IFSS) and Approved Document B (Fire Safety) of the Building Regulations.

How Can You Demonstrate Levels 1 and 2?

Your assessors will start at the highest level you have declared, which will be level 2.

At level 2, RICS expect that you will be able to provide practical experience in some of the following activities or tasks:

- Practical application of health and safety issues, including compliance and legislative requirements;
- Obtaining formal qualifications, such as first aid, nationally recognised and industry-specific qualifications;
- Acting in a specific role under key legislation, such as CDM 2015;
- Being involved in a health and safety audit or review;
- Reviewing health and safety proposals or risk assessments and method statements (RAMS) as part of a tender or tender analysis.

Typical questions will start with, 'How have you safely inspected . . . ?', 'Tell me about an example of when you carried out a risk assessment . . .' or 'How did you ensure you were safe on-site when . . . ?'. Remember, all the assessors' questions should be based upon the examples you provide in your level 2 write-up for this competency in your summary of experience – rather than being hypothetical or general.

You may also be asked some level 1 questions, focusing on the knowledge behind your practical level 2 examples.

Chapter 6

Accounting Principles and Procedures

Accounting Principles and Procedures is a mandatory RICS APC competency required to level 1. This means that you need to write up your summary of experience with a statement of knowledge at level 1 only.

The RICS pathway guide sets out this competency's relevant knowledge (level 1). This covers basic accounting principles and interpretation of company accounts. This is an essential skill for surveyors when working in or running a business, vetting suppliers or contractors or considering tenant covenant strength in a valuation instruction. Understanding a set of accounts or a company's financial performance is also a valuable life skill.

This chapter will explain what you should know concerning Accounting Principles and Procedures, covering the main requirements of level 1.

Remember, as a surveyor, you will have limitations regarding your depth and breadth of accounting knowledge. Always consult or refer a client to an accounting or taxation specialist if in doubt.

Company Accounts

Understanding company accounts is a crucial accounting skill at level 1. Company accounts are generally either in the form of financial (statutory) or management accounts.

As a starting point, here are some key accounting terms to be aware of:

- Accrual – money that a business has earned or spent but has not yet been paid, e.g. a payment from a client or an amount owed to a supplier;

DOI: 10.1201/9781003437116-6

- Turnover (revenue) – total sales;
- Cost of sales – the cost of producing the goods and services a business sells;
- Gross profit – turnover minus the cost of sales, e.g. wages relating to a specific service;
- Gross profit margin – gross profit divided by revenue, multiplied by 100;
- Operating expenses – the expenditure needed to support the overall running of the business, such as rent, rates and utility costs;
- Net profit (bottom line) – gross profit minus operating expenses (overheads) and taxes. This is the amount of money that can be distributed to the business owners or reinvested in the business;
- Net profit margin – net profit divided by revenue, multiplied by 100. A high profit margin is desirable to a business.

Financial accounts are required by law, e.g. under the Companies Act 2006. They are prepared annually in a fixed statutory format, following a specific accounting convention. Management accounts are far more flexible, are prepared more regularly and are for internal management use only.

Company accounts can generally be downloaded from Companies House. We recommend downloading two sets of accounts to reflect upon whilst you are reading the rest of this chapter. Try to find one set for a small surveying business set up as a limited company and one for a large surveying business set up as a PLC. You could also download your own firm's accounts to understand more about where you work. If your firm has a separate accounting team, spending time with them to understand accounting in further detail is a great way to expand your knowledge.

The UK tax year runs from 6th April to 5th April the following year. However, a company's financial year will usually be for 12 months starting on the day that the company was formed, although this can be amended in certain circumstances.

There are three key financial statements:

- Balance sheet/statement of financial position – this provides a view of a business's financial position, i.e., assets = equity + liabilities, on a given date. This means that the balance sheet provides a snapshot in time;

- Income statement/profit & loss statement – this summarises a company's income and expenditure over the accounting period. This provides an overview of company performance, including the net profit or loss. This can be compared to other accounting periods;
- Cashflow statement – this merges the balance sheet and income statement to show actual receipts and expenditure, including VAT. The cashflow statement is split into core, investing and financing activities and demonstrates how (and why) the business' bank balance changed over the accounting period.

A qualified accountant should prepare a company's financial accounts. The business should provide their account with their accounting records from the year to allow the accountant to do this. Businesses will typically use accounting software, such as Xero, Sage or Quick-Books, and potentially a bookkeeper to do this.

Companies House states that private companies must keep their accounting records for three years and PLCs for six years.

Under the Companies Act 2006, a company's financial accounts may need to be audited by an accountant to confirm that there are no material misstatements. All PLCs must have their accounts audited, whilst some smaller companies may claim an audit exemption based on their size.

You may need to review a company's accounts and/or a credit report as part of the supplier or contractor vetting process or when analysing a tenant's covenant in a valuation instruction. The activity itself is outside the scope of level 1 (knowledge only). However, a basic understanding of what you might find in a credit report is helpful and relevant level 1 knowledge.

Various companies publish credit reports, including Dun & Bradstreet, Creditsafe and Experian. They are all relatively similar, including:

- A credit rating, typically from A–E with A being the strongest, alongside an associated numerical rating;
- Information on the business;
- Ownership;
- Key financials, including a summary and in-depth data.

Companies House Requirements

All UK private limited companies, LLPs and PLCs must comply with the Companies Act 2006, including filing annual accounts with Companies House.

Companies are split into size categories for accounting purposes: micro-entity, small, medium and large. These are based on turnover, balance sheet and employee numbers. This is a complex area of accounting practice, and a specialist should always be consulted for confirmation.

Micro-entities and small companies can prepare abridged accounts, with a more straightforward form balance sheet and other minor differences to full accounts.

Parent companies usually need to prepare group accounts.

Dormant companies must also submit accounts, although the requirements are far more straightforward.

Accounts must be filed within nine months of the year-end for private companies and six months for PLCs. Failing to file or file company accounts late is a criminal offence and financial penalties also apply to late filing.

We have summarised the essential requirements for company accounting in the UK next, although a full in-depth summary can be found on the Companies House website (Companies House, 2023).

A set of company accounts should include the following:

- Profit and loss statement;
- Balance sheet;
- Notes to the accounts.

Depending on the company's size, the accounts may need to be accompanied by a directors' report (e.g. for a medium size and above company) and an auditor's report.

The accounts must be physically signed by a director or company secretary, the accountant who prepared the accounts and the auditor (if relevant).

Companies must also file an annual confirmation statement to Companies House. This confirms vital information about a company, such as directors, secretaries, people with significant control, registered office address, shareholder information and Standard Industrial Classification (SIC) code.

Financial Reporting Standards

In the UK, companies must adopt one of two financial reporting standards to comply with the Companies Act 2006. These are the UK Generally Accepted Accounting Practice (GAAP) and the International Financial Reporting Standards (IFRS).

IFRS must always be adopted by PLCs in their group accounts. Whilst other companies may have a choice of which to adopt, UK GAAP is the most commonly applied standard. This is because its requirements are more straightforward and more flexible, and the accounts are less comprehensive than under IFRS.

Under UK GAAP, the three key Financial Reporting Standards (FRS) are 101 (reduced disclosure framework), 102 (small and medium entities) and 105 (micro-entities). An accounting specialist should be consulted about which FRS should be adopted for a specific company.

The RICS Valuation – Global Standards 2017: UK National Supplement (effective January 2019) includes guidance on valuations for financial reporting in UK VPGA 1. This confirms that under FRS 102, assets, such as property, should be measured based on the cost, revaluation or fair value model. Candidates with Valuation as a technical competency and who provide valuations for financial reporting purposes should read this UK VPGA in full.

One final issue to be aware of is IFRS 16, which relates to lease accounting. It effectively brings all leases onto the balance sheet so that they are treated as finance leases instead. This makes it easier to assess an entity's lease commitments and compare different entities using their financial statements.

Under IFRS 16 (IFRS, 2019), a lease is defined as a contract that 'conveys the right to control the use of an identified asset for a period of time in exchange for consideration'. Leases under 12 months in duration are excluded. IFRS 16 impacts various financial metrics, such as gearing ratio and EBITDA. This means that specialist accounting advice will be needed.

How Can You Demonstrate Level 1?

Your assessors will start at the highest level you have declared, which will be level 1. This means that you only need to state your knowledge (level 1) and do not need to include any examples for this competency in your summary of experience.

Typical questions will start with, 'Tell me about your knowledge of . . .', 'What do you know about . . . ?' or 'Explain . . .'. You can substitute any of the accounting principles and procedures outlined in this chapter to write yourself a set of sample questions.

Reference List

Companies House, 2023. *Accounts Guidance*. [Online] Available at: www.gov.uk/government/publications/life-of-a-company-annual-requirements/life-of-a-company-part-1-accounts [Accessed 18 July 2023].

IFRS, 2019. *IFRS 16 Leases*. [Online] Available at: www.ifrs.org/issued-standards/list-of-standards/ifrs-16-leases/ [Accessed 18 July 2023].

Chapter 7

Business Planning

Business Planning is a mandatory RICS APC competency required to level 1. This means that you need to write up your summary of experience with a statement of knowledge at level 1 only.

The RICS pathway guide sets out this competency's relevant knowledge (level 1). Business planning refers to defining a business's future, usually through a written plan.

This chapter will explain what you should know concerning Business Planning, covering the main requirements of level 1.

Business Plans

A business plan is a written document defining a business's future objectives and strategy, i.e., what it wants to achieve and how it will do this.

Setting out the plan in writing is essential as it ensures the business is accountable and committed to planning for the future. It also means that the plan can be reviewed and reflected upon at various points in the future. Without a business plan, a business is potentially jeopardising its future success.

A business plan is not a static document – it should be reviewed regularly (such as annually) and updated to take account of market changes, business changes, amended objectives or new strategies. Setting a date in the future to review the business plan is good practice.

There are many types of business plans, although the RICS mentions strategic, departmental, operational and corporate in the pathway guide. These all have different time horizons and objectives. A business may have multiple business plans, such as an overall top-level strategic plan, an operational plan covering various business departments and

DOI: 10.1201/9781003437116-7

separate department-level business plans. A corporate business plan is similar to a strategic plan but is generally used where a group company has a complex structure with various subsidiaries or business units.

A business may also have a contingency business plan, setting out what will happen in various unforeseen scenarios. This could include a key staff member being unavailable, IT systems being down or a reputational issue. This plan will set practical steps for the business to recover from the situation and continue trading.

Business plans can also be used for a variety of purposes, such as:

- Raising funding or finance;
- Gaining new instructions, clients or customers;
- Focusing on priorities;
- Responding to change;
- Budgeting or managing financial resources;
- Setting staff targets.

If a business needs to raise finance or investment, there are several different sources to consider:

- Bank finance (such as loans or mortgage lending if the business owns the property);
- Peer-to-peer lending;
- Crowdfunding;
- Equity finance (selling a stake or shares in the business);
- Government grants or loans;
- Own funds;
- Venture capital.

Various key elements can be found in a business plan (although this is not an extensive list as the contents will be specific to the plan in question):

- Executive summary;
- Mission statement;
- Vision statement;
- Description of business opportunity or idea;
- Team;
- Operations;
- Marketing;

- Sales;
- Competitive analysis;
- Forecasted cashflows;
- Objectives and goals.

A mission statement will define and focus on the business and its objectives today, whereas a vision statement will consider these in the future. The mission statement will drive the company, whereas the vision statement will direct the company.

For example, the mission statement of the RICS could be said to 'promote and enforce the highest professional standards in the development and management of land, real estate, construction and infrastructure' (RICS, 2023).

The goals set out in a business plan could adopt the SMART model. This means that goals are specific, measurable, achievable, relevant and time-bound.

Business planning is also likely to include market research. This is where you identify a gap in the market, a new market or demand for a specific service. Critical steps in this process include determining the following:

- Target audience;
- Competition;
- Problem or challenge to be solved;
- Current strengths and weaknesses;
- Pricing;
- Branding;
- Marketing.

A SWOT analysis is an excellent tool for this process, considering strengths, weaknesses, opportunities and threats. An example SWOT analysing a new surveying business is set out in Figure 7.1.

PESTLE analysis could also be used, which considers external factors affecting a business. These are split into political (e.g. Brexit), economic (e.g. inflation), sociological (e.g. demographics), technological (e.g. BIM or e-conveyancing), legal (e.g. new break clause case law) and environmental (the Minimum Energy Efficiency Standard, or MEES).

Understanding basic economics is an integral part of business planning. This includes terms such as demand, supply, market and industry.

Figure 7.1 Example SWOT Analysis

Reading a good university-level textbook is an excellent way to acquire relevant knowledge, e.g., Economics and Property by Danny Myers.

Economics is vital in understanding the market for a firm's services (i.e., demand from the market) and who the firm is competing with (i.e., close substitutes in the industry) in terms of supply.

Porter's Five Forces Model can be used to analyse competition for a firm's services. This means considering the following in terms of a surveying business:

- Competition and their service offering;
- Potential new entrants who could increase the supply;
- Supplier power and impact on the cost of inputs;
- Customer power and effect on the cost of services;
- The threat of substitutions that could be purchased instead of the firm's services.

Surveyors should have read their firm's business plan and know how they work towards it in their day-to-day work. You can also find business plan templates on the Prince's Trust website.

A good business plan should be:

- Short, concise and to the point;
- Regularly updated;
- Inclusive of relevant information, including additional data, in appendices if appropriate;

- Realistic, not overly optimistic or based on inaccurate financials or data;
- A formal, professional document with a cover, contents and visuals;
- Reviewed and updated regularly.

All surveyors should understand the current RICS Business Plan. This is regularly updated, so we recommend downloading the current copy from the RICS website.

Organisational Structures

Firms come in various shapes and sizes, from the most prominent international firms to sole traders. Each will have adopted an organisational structure and business type that suits their needs.

At the top level, organisational structure relates to several factors:

- Whether the business is centralised (with decisions made by senior management and disseminated to employees) or decentralised (where decisions are made by employees and then communicated to senior management). Each has its strengths and weaknesses, particularly concerning the time to make decisions and the level of staff autonomy;
- Physical structure, such as functional (departments based on work specialism), divisional (groups based on shared services, products or locations), matrix (mixed specialism teams with separate managers), team (with each delivering one product or service), network (joining together multiple businesses or contractors), hierarchical (direct chain of command with various levels of seniority) and flat (typical in small companies with fewer employees where seniority is not defined between employees).

Find out what structure your business uses and look at the reasons why this might be the case. You could use a SWOT analysis to reflect upon whether your firm's organisational structure meets the needs of its business plan.

UK businesses come in various operating structures, such as sole trader, partnership, limited liability partnership and limited company. Each has its requirements relating to personal liability and tax.

The Companies Act 2006 is the key legislation relating to company law in the UK, although in-depth knowledge of this will be outside the scope of level 1. Accounting requirements will be discussed in this book's Accounting Principles and Procedures chapter.

We will discuss each, with an overview only given on tax requirements. If you are in any doubt or would like to know more about the tax implications of a given business structure, we recommend that you speak to an accountant or tax advisor. Also, note that tax rates frequently change, so these should always be checked on the HMRC website.

A sole trader is a straightforward form of business where a surveyor is self-employed and runs the company as an individual. This is usually common for start-ups or micro-surveying businesses. The surveyor is effectively the business for legal purposes, so they will be personally liable for all company liabilities (i.e., debts). This can be risky as a surveyor may need to use personal assets to pay the business's debts, e.g., their house. The sole trader does not have to register with Companies House. However, they do have to inform HMRC as they will retain the business' profits after Income Tax is paid through self-assessment.

A limited company separates the business from the owners (or directors), which effectively 'ring fences' the business from personal assets and finances. Limited companies must be registered at Companies House, with a SIC code identifying the company's specific business. Directors of limited companies have a range of reporting and management responsibilities, including following the company's rules (as per the articles of association), good record keeping, filing accounts and paying Corporation Tax (on the company's profits, investments and chargeable gains when an asset is sold).

A Public Limited Company (PLC) is similar to a limited company but can sell its shares to the public on a listed stock exchange, such as the London Stock Exchange (LSE). PLCs have additional reporting requirements and responsibilities to be accountable to their shareholders.

Two or more sole traders could form a partnership, similar to being a sole trader. The partnership must be registered with HMRC so each partner pays Income Tax through self-assessment. Again, each partner is personally liable for the losses of the business, and if one partner cannot pay, the other will be liable for their share. This makes a partnership a potentially risky way of entering into business with someone else.

Businesses can also be set up as a Limited Liability Partnership (LLP), a much less risky way of entering into business with others as the business is effectively 'ring fenced' from each partner's personal assets and finances. An LLP must be registered with Companies House and HMRC, and there is a similar level of administration and reporting to a limited company. However, the partners are assessed for Income Tax individually through self-assessment rather than the business paying Corporation Tax.

Businesses must also consider whether they need to register for VAT, as this can affect their pricing, cashflow and reporting requirements. VAT is a tax added to specific products and services where a business is registered for VAT. A company must register for VAT where their turnover exceeds a threshold, which in 2023 was £85,000. Businesses can voluntarily register for VAT if their turnover is below this threshold.

When a business registers for VAT, it must charge VAT on all goods and services sold, and it can reclaim the same on business expenditure. VAT returns are required to be submitted to HMRC online every three months.

VAT is charged at three rates:

- 20% – most goods and services;
- 5% – reduced rate goods and services;
- 0% – zero rate goods and services, such as food, train fares and books.

Some goods and services are exempt from VAT, such as postage stamps.

Surveyors working in construction must also know about the Construction Industry Scheme (CIS) and VAT domestic reverse charge.

Surveyors setting up in business must consider the RICS requirements to register as a Regulated Firm. These are discussed in detail in this book's Ethics, Rules of Conduct and Professionalism chapter.

Financial Benchmarking

Effective business planning requires an in-depth understanding of the financial data underlying business performance. This allows the business to understand how their performance compares to competing firms in the market.

Generally, financial benchmarking will focus on several key financial ratios or benchmarks. Before we look at these, we will define some key terms:

- Liabilities – debts that a business owes;
- Current liabilities – debts that a business owes and must pay within 12 months, e.g., VAT bill, wages owed or credit card debt;
- Assets – what a business owns;
- Current assets – what a business owns and expects to use or sell within 12 months, e.g. inventory, cash in the bank or invoices owed by clients;
- Stock – the value of goods that a business has to sell to customers. If a company solely sells services, then it will not own any stock;
- Debtor – who owes the business money;
- Creditor – who the business owes money to.

Key financial ratios to be aware of include:

- Working capital ratio – the difference between current assets and current liabilities, or the ability of a business to cover its outgoings in the short term. This ratio should be positive, showing that the business can meet its liabilities and may have a surplus to invest in the business;
- Quick ratio (acid test) – this is calculated by dividing liquid assets (i.e., cash or close to cash) by current liabilities. It shows that a company can meet current liabilities with assets that can be converted quickly to cash. This ratio can be compared between companies and used as a snapshot of financial health;
- Debt to equity ratio – this is calculated by dividing a company's total liabilities by total shareholder equity. It shows a business's reliance on debt finance, such as loans, otherwise known as leverage;
- Return on equity – this is calculated by dividing net income by average shareholder equity. It measures a business's profitability and efficiency in generating profits. Generally, a higher ratio shows that a business is better at converting equity into profit – which is great for the shareholders of a business!

Many other financial ratios could be used, depending on the business's specific benchmarking and performance measurement requirements.

Accounting and Forecasting Techniques

Forecasting is an essential part of business planning. It should be included within the business plan and updated regularly over a suitable time horizon, e.g. 12 months.

Forecasting will only be as accurate as the data used. This means that businesses should keep accurate financial records if they want to be able to forecast effectively. Reasonable assumptions about future income and expenditure may need to be made to facilitate forecasting.

Many surveyors will be involved in planning and forecasting their workload, generally using a work-in-progress database, bespoke web-based system or a simple spreadsheet. Regularly updating this data is essential to ensure that forecasting is as accurate as possible.

There are two types of forecasting; qualitative and quantitative. Quantitative forecasting is based on historical data, whereas qualitative methods use experience and knowledge of the business and market. Using both can produce the most valuable and accurate forecasts for a business.

For construction professionals, RICS publishes a Guidance Note called Cash Flow Forecasting (1st Edition). This is part of the Black Book and relates specifically to cash flows for construction projects rather than for businesses. If relevant, we recommend that candidates undertake further reading into this Guidance Note, which will also be discussed in further detail in Routledge's series of pathway-specific books, focusing on the technical competencies.

How Can You Demonstrate Level 1?

Your assessors will start at the highest level you have declared, which will be level 1. This means that you only need to state your knowledge (level 1) and do not need to include any examples for this competency in your summary of experience.

A beneficial exercise is to read your firm's business plan and understand how this relates to your work. You can then mention this in your level 1 write-up. We also recommend reading the RICS business plan, mainly as so much change has happened over the past few years given the context of Covid and the Levitt and Bichard Reviews.

Typical questions will start with, 'Tell me about your knowledge of . . .', 'What do you know about . . . ?' or 'Explain . . .'. If you can link your responses to your work and experience, e.g., mentioning your firm's business plan, this will help show that you know and understand business planning appropriately.

Reference List

RICS, 2023. *About RICS*. [Online] Available at: www.rics.org/about-rics [Accessed July 27 2023].

Chapter 8

Conflict Avoidance, Management and Dispute Resolution Procedures

Conflict Avoidance, Management and Dispute Resolution Procedures is a mandatory RICS APC competency required to level 1. This means that you need to write up your summary of experience with a statement of knowledge at level 1 only.

The RICS pathway guide sets out this competency's relevant knowledge (level 1). This includes:

- Common causes of disputes pertinent to your area of practice;
- Ways to avoid conflicts, such as risk management, early warning signs, partnering techniques and clear and robust client briefs;
- Alternative Dispute Resolution (ADR) methods;
- Theories of negotiation and the role of effective communication and negotiation (covered in the Communication and Negotiation chapter of this book);
- The roles of an expert witness and an advocate;
- The role of the RICS Dispute Resolution Service (DRS);
- Conflicts of interest and the related RICS Professional Standard (covered in this book's Ethics, Rules of Conduct and Professionalism chapter).

This chapter will explain what you should know concerning Conflict Avoidance, Management and Dispute Resolution Procedures, covering the main requirements of level 1.

The RICS publishes comprehensive guidance concerning this competency. The primary guidance is the Guidance Note Conflict Avoidance and Dispute Resolution in Construction (1st Edition). Supplementary guidance will be discussed in the individual sections below.

DOI: 10.1201/9781003437116-8

Conflict Avoidance

There are many common causes of disputes in surveying, including money (e.g. unpaid rent, construction contract payments and outstanding invoices), delays, defects, differing objectives, personality clashes and differences of opinion. You need to be aware of those relevant to your experience and area of practice.

Of course, it is always advisable to try to avoid conflict or disputes before they occur. There are many ways to do this, including:

• Risk management – identifying the causes of disputes to minimise or prevent them;
• Early warning signs – this means identifying early symptoms of problems, such as a breakdown of communication, an increase in written notices or letters or late payments;
• Partnering and alliancing techniques – this is where key stakeholders make a joint agreement or commitment at the outset to pursue mutual goals and objectives. Cooperation helps projects or instructions to run more smoothly;
• Good payment practices, reporting and record-keeping during projects;
• Precise and robust client briefings or contract documentation.

Alternative Dispute Resolution (ADR) Methods

Dispute resolution is a simple process to resolve a conflict between two parties. Generally, this will be litigation in court (following the Civil Procedures Rules, or CPRs) or using ADR (i.e., not resorting to litigation).

Under the Pre-Action Protocols to the CPRs, the parties are encouraged to use ADR and to reach a settlement outside of court. This aims to make the civil justice system simpler, quicker and cheaper, to prevent the court system from being used as a negotiation tactic and to ensure that the cases are dealt with proportionally.

ADR is typically cheaper, more flexible and quicker than litigation. It can also be confidential and helps preserve the parties' relationship.

The main dispute resolution techniques have been split into three pillars (Mackie, et al., 2007): negotiation, mediation or conciliation, and adjudication.

However, it is perhaps simpler to work through the main ADR processes, which are mediation, adjudication, arbitration and expert determination.

Mediation is where a neutral third party mediator facilitates discussions and helps the parties reach a settlement that can be legally binding if documented in a written contract. Mediation is a confidential and informal process, usually undertaken without prejudice. Even if a contract does not mention mediation, it could be used by the parties as an alternative to litigation or specified ADR methods in the contract.

Adjudication is a common method of resolving construction disputes under Section 108 of Part 2 of the Housing Grants, Construction and Regeneration Act 1996 (amended by the Local Democracy, Economic Development and Construction Act 2009). Adjudication applies to construction contracts falling under Section 104 of the Act and can be used to resolve a dispute between two parties within 28 to 42 days. It produces an interim binding decision, later determined by another process, such as litigation, arbitration or negotiation. Adjudication relies on the principle of pay first, argue later and can help to preserve cashflow during construction.

Arbitration is governed by the Arbitration Act 1996 and is often used in commercial rent review disputes, where the revised rent cannot be agreed upon by negotiation. The lease will usually specify whether the third party acts as an arbitrator or expert.

A third party arbitrator receives written representations and replies from both parties' representatives (acting as expert witnesses) and makes a legally binding award. The arbitrator has the power to award all costs, and there are limited grounds of appeal against an arbitrator's award.

Expert determination is another common way of resolving rent review disputes. A common misunderstanding among APC candidates is using the terms 'independent expert' and 'expert witness'. An independent expert relates to the role of the third party during ADR proceedings. In contrast, an expert witness is a role that a surveyor (representing their client, such as the landlord or tenant) may act in during third party proceedings. They are very different roles, and a clear understanding is required of both. During expert determination, an expert provides their expert opinion to determine the matter. Depending on the contract or lease terms, the expert may or may not be required to consider written representations and replies put forward by each party's representative (i.e., expert witness). The expert may be required to give a reasoned determination and will have power over costs as determined by the lease or contract. Unlike an arbitrator, an expert can be liable for negligence, and their determination is not legally binding.

For commercial lease renewal disputes, RICS runs a service called Professional Arbitration on Court Terms (PACT), which can involve either a third party arbitrator or expert. PACT offers a viable alternative to court where the parties cannot agree on key lease terms, such as the new lease term, rent, interim rent, service charge, repair, alienation or break clauses.

Dispute Review Boards (DRB) and Dispute Adjudication Boards (DAB) are also terms that you may come across concerning construction disputes, particularly in relation to FIDIC standard forms of contract (Clause 20 refers to the FIDIC DAB). DRBs make recommendations, whereas DABs make binding decisions on disputes.

JCT contracts, on the other hand, typically refer to mediation, adjudication and arbitration or litigation in court. This will be covered in the pathway-specific series of books.

The Role of the Expert Witness and Advocate

We have already touched upon the role of an expert witness, a role that a surveyor may act as during third party proceedings. The role is clarified in the RICS Practice Statement and Guidance Note Surveyors Acting as Expert Witnesses (4th Edition). Part 35 of the CPRs relates to the role of the expert witness, who has a primary duty of care to the court and should remain uninfluenced by their client. An expert witness must sign a statement of truth before giving expert witness evidence, and they cannot act on an incentive or contingency fee basis. The expert witness must provide independent unbiased advice within their experience and knowledge, stating the main facts and assumptions they have made in doing so (and without omitting material facts relevant to the conclusions they have made).

Historically, an expert witness was immune from being sued for dishonest or negligent statements. However, this changed following the case of *Jones v Kaney* [2011] UKSC 13 where an expert witness can be found liable for a breach of their duty of care to their client in contract or negligence (as per the case of *Hedley Byrne & Co Ltd v Heller & Partners Ltd* [1964] AC 465). In simple terms, an expert witness should only accept instructions where they are competent and can justify their conclusions with robust evidence.

In contrast to the role of an expert witness, an advocate represents their client in third party proceedings or court. They have a duty to the tribunal or court to act properly and fairly. However, they also have a duty to act in their client's best interests. An advocate still needs to be competent to act, although they are not obligated to disclose all material facts or matters they are aware of or that affect their opinion.

In some circumstances in Scotland, usually lower-order tribunals, surveyors may be permitted to act in a dual expert witness and advocate role. Where relevant, you can read more about this in the RICS Professional Statement and Guidance Note Surveyors Acting as Expert Witnesses in Scotland (1st Edition).

RICS Dispute Resolution Service (DRS)

The RICS Dispute Resolution Service (DRS) provides for the appointment of a variety of third party dispute resolvers, including arbitrators and experts for commercial property rent reviews (using a DRS1 form), adjudicators, mediators, ground rent assessments, neighbourhood disputes and simplified arbitration services for rural rent reviews.

In rent review cases, the lease will confirm who can make the appointment, and a fee will be payable to RICS at the point of application. Surveyors dealing with rent review or lease renewal disputes should always consider putting costs at risk (including the costs of any third party application) by using a Calderbank (rent review or lease renewal) or Part 36 offer to settle (lease renewals only).

RICS also publishes extensive guidance for dispute resolvers, which is available in the Dispute Resolution Standards section of the RICS website.

How Can You Demonstrate Level 1?

Your assessors will start at the highest level you have declared, which will be level 1. This means that you only need to state your knowledge (level 1) and do not need to include any examples for this competency in your summary of experience.

Typical questions will start with, 'Tell me about your knowledge of . . .', 'What do you know about . . . ?' or 'Explain . . .'. You can substitute any of the topics outlined in this chapter to write yourself a set of sample questions.

Reference List

Hedley Byrne & Co Ltd v Heller & Partners Ltd AC 465 (1964).
Jones v Kaney UKSC 13 (2011).
Mackie, K., Marsh, W. & Miles, D., 2007. *Commercial Dispute Resolution: An ADR Practice Guide*. London: Butterworths.

Chapter 9

Data Management

Data Management is a mandatory RICS APC competency required to level 1. This means that you need to write up your summary of experience with a statement of knowledge at level 1 only.

The RICS pathway guide sets out this competency's relevant knowledge (level 1). This covers the collection, collation, analysis, storage and management of information sources and data relevant to your area of surveying practice. It also includes understanding applicable legislation, such as the Data Protection Act 2018, and how data are centrally stored (usually online) for projects or instructions.

This chapter will explain what you should know concerning Data Management, covering the main requirements of level 1.

Data Sources

You will likely work with various data and information sources during your career. These will depend on your practice area and the specific instruction or project you are working on.

All data sources need to be:

- Verified;
- Accurate;
- Up to date;
- Reliable.

Examples for a typical valuation instruction could be:

- British Geological Survey for soil type;
- Coal Authority for historic coal mining activity;

DOI: 10.1201/9781003437116-9

- Environment Agency for flood risk;
- Land Registry for tenure;
- Rightmove, CoStar or EIG (amongst many other online databases) for comparable evidence research;
- Valuation Office Agency (VOA) website for business rates.

Having a checklist or template for the typical data sources you would interrogate is helpful to ensure that you do not miss any out. Some instructions will have specific issues you must research or find data on, e.g., non-traditional construction, flood risk or quality of retail pitch.

When researching comparable data, you need to ensure that your comparable evidence is:

- Comprehensive, i.e., ideally, more than one transaction. At least three, if not more than five, would be better;
- Very similar or identical;
- Recent;
- Result of an arm's length transaction;
- Verifiable;
- Consistent with local market practice;
- Result of underlying demand, i.e., sufficient bidders to create an active market.

By way of an example, when researching comparable rental evidence for lease renewals and rent reviews, you could adopt the following process:

- Use published online databases, such as CoStar or EIG, to identify potential comparables;
- Look for done deals, such as lettings, lease renewals, rent reviews and regears. Consider the weight allocated to each under the hierarchy of evidence (i.e., to what extent is the transaction at arm's length?);
- Look for investment or auction sales (completed deals or properties being marketed), where you can extract details of key leasing events. Some may be historical, so ensure you keep a close eye on any transaction dates;
- Look for properties that are available to let. Although not done deals, these may indicate ceiling (quoting) rents and the market dynamics in a specific location. Speaking to the agents involved may give you further leads on completed deals nearby;

- Speak to active local and national agents in the area. Some online databases have a search function to identify which agents have recently done deals in an area. You are likely to locate evidence not published on the online databases this way, including rent reviews, regears and off-market transactions;
- Use an online database (such as The Requirement List) to identify active requirements in the area and the levels of rent that tenants might pay;
- Look at nearby properties in the area that could be potential comparables and use the Land Registry to identify landlord or tenant contacts or to download available leases.

You must speak to the agent involved with each comparable identified to verify the facts. Ideally, speak to both parties as they often have different analyses or views.

As part of your discussions and due diligence, you should record the following details:

- Address;
- Type of property;
- Tenure, i.e., freehold or leasehold;
- Location;
- Lease terms, if applicable;
- Brief description;
- Floor areas;
- Transaction type and date;
- Value and any incentives;
- Analysis;
- Parties involved;
- Source of information;
- Any issues relating to reliability;
- Date the evidence was verified;
- Having a simple template proforma to record key details is best practice. This can then be saved on your file for future reference.

You will then need to analyse your comparable evidence:

- Analyse the net effective rate, i.e., taking any incentives into account;
- Collate evidence in a schedule or matrix;

- Adopt a common measurement or comparison standard;
- Adjust quantitatively and qualitatively using the hierarchy of evidence and other vital considerations (including your knowledge and experience);
- Analyse the evidence to form your opinion of value;
- Stand back and look to sense-check the final valuation figure;
- Report your opinion to the client, including any negotiation strategy or quoting terms.

You can apply these data management principles to any instruction or project.

Candidates will undoubtedly be aware of artificial intelligence (AI) tools, such as ChatGPT. Under no circumstances should you attempt to use these to write your submissions or to write unedited copy for instructions, reports or projects. AI might be helpful for preliminary desktop research, but it is no substitute for the knowledge and human intelligence that a surveyor brings to their work. There is also an inherent plagiarism risk with the use of AI tools, which will be picked up by Turnitin when written work is submitted through the RICS Assessment Platform.

Data Security

The data you collect must be stored securely within your firm's IT system. You should read your firm's IT and data management policies and be ready to discuss these in your interview.

Ways that you can keep data secure include:

- Firewalls;
- Passwords being changed every 30 days or as per your firm's IT policy;
- Not leaving devices unattended or visible in your car or on your desk;
- Using discrete departmental drives to avoid wider sharing of information (thus helping to prevent or manage conflicts of interest between departments);
- Encryption;
- Virus protection (including on mobile and Apple devices);
- Not opening and alerting your IT team to suspicious email attachments or links;

- Being cautious about sharing on social media;
- Backing up data;
- Applying automatic updates, which may provide additional security features;
- Using two-step verification.

There are a variety of data security threats that you may encounter:

- Ransomware – this is malware that encrypts and threatens to destroy, remove access to or publicly post data unless a victim makes a payment, which often increases as time elapses;
- Phishing and whaling (also known as CEO fraud) – this is a malicious attempt to acquire sensitive information by masquerading as a trustworthy source via email, text or pop-up message or to coerce someone into making a money transfer;
- The exploitation of software vulnerabilities – these are flaws, glitches or weaknesses discovered in software;
- Insider threat – this is where an employee causes a data breach, either by mistake or on purpose;
- Hacking;
- Loss or theft of equipment.

Electronic signatures are also becoming more popular, with the Land Registry accepting these in certain circumstances. There are various data security issues when using electronic signatures, so only secure forms should be used.

Data security is also a key consideration when managing a conflict of interest using an information barrier. RICS defines an information barrier in their Professional Statement Conflicts of Interest (1st Edition) (RICS, 2017) as 'the physical and/or electronic separation of individuals (or groups of individuals) within the same firm that prevents confidential information passing between them'. In the Ethics, Rules of Conduct and Professionalism chapter, you can read more about conflicts of interest.

An information barrier could be used where different departments in the same (usually a larger) firm are acting for two (or more) clients on the same property. All clients must have given written informed consent to manage the conflict of interest using an information barrier. The information barrier must prevent any communication of information relevant to the instruction between the various departments, including any administrative resource. The firm will need to ensure that access and data security protocols are implemented in relation to both physical and electronic files, typing resources, printers, access to email and other data sources. In practice, this could include locked filing cupboards, password-protected access and separate administrative resources for each department.

Social Media

Social media has become increasingly popular, with the property and construction industry adopting it as the norm. Popular social media platforms include LinkedIn, Twitter, Facebook, Instagram, TikTok, forums, podcasts and blogs. However, different platforms will have different purposes and uses, with some being more appropriate for business than others.

LinkedIn, in particular, is an excellent business tool that can provide a visible online CV for surveyors and the opportunity to connect with other professionals, clients and key stakeholders.

Social media has many benefits to businesses, such as networking, information sharing, business generation, marketing, advertising, raising awareness, reaching a wider audience and collating reviews and recommendations.

RICS published a Regulation Paper on the Use of Social Media: Guidance for RICS Members (Version 1). This is because the line between surveyors' personal and professional lives often becomes blurred online.

The RICS Rules of Conduct require surveyors to promote trust in the profession and treat others respectfully. This applies to our online presence and communications, not just face-to-face or traditional media, such as emails, letters and meetings. This does not mean that surveyors should not have an online presence and use this to express their opinions, messages and values. However, we need to take responsibility for our online presence and ensure that we remain professional and respectful when we post.

RICS may investigate a social media post that adversely impacts public confidence or trust in the profession. This includes posts that:

- Are discriminatory (which would also breach the Equality Act 2010), dishonest, abusive or threatening;
- Bully, harass or victimise another person or people;

- Show a pattern of frequent or a high number of concerning posts;
- Ignore previous advice or warnings about RICS concerns or a request from RICS to remove a post.

RICS will not investigate posts that are critical of an organisation or its policies and performance, provided that they use professional and respectful language. However, if the criticism is of an individual, they may investigate.

Social media posts by surveyors on their personal accounts may fall within the remit of RICS Regulation if they are extreme or highly offensive. Furthermore, approval for, or sharing others', discriminatory posts may also lead to an investigation by RICS.

Following an investigation, RICS may take disciplinary action in serious cases.

Data Systems and Databases, Including Building Information Modelling (BIM) and Technical Libraries

An Electronic Document Management System (EDMS) is software (usually a Cloud/online database) which collectively stores comprehensive data and documents. This is an alternative to paper files or a simple electronic folder-based system.

EDMS relies on sound data management principles being followed, such as:

- Controlling access using user accounts and passwords;
- Having naming conventions for documents;
- Keeping a record of distributed and disposed data;
- Having user procedures in place;
- Having process sheets, e.g. for setting up new files or folders;
- Having processes to deal with data redundancy, where the same piece of data is stored in multiple locations.

Firms may also comply with ISO 27001 (information security) or 9001 (quality management) in relation to their IT and/or EDMS systems.

If you work in the construction industry, you may come across Building Information Modelling (BIM). This will be discussed in depth in the relevant pathway-specific books. However, large amounts of data will be processed during a project's lifecycle, including structured data (e.g. schedules or a BIM model) and unstructured data (e.g. emails, photographs or meeting notes). If you are using BIM, you will need to understand how BIM documentation and data are stored within models and software.

You may also come across technical libraries in various shapes and sizes that store different data and documents, including product information, journals, articles and books.

Blockchain is a reasonably new technology that has increased use in the property industry. In simple terms, Blockchain can automate transactions, create smart contracts and store all information about an asset, including ownership.

Key Legislation

All surveyors need an excellent working knowledge of the Data Protection Act 2018 and the UK General Data Protection Regulation (GDPR), the primary legislation affecting data management in the UK. These repealed the former Data Protection Act 1998 and originally implemented the EU General Data Protection Regulation (GDPR) (before Brexit).

The 2018 Act applies to data controllers (who determine how and why personal data is processed) and processors (who act on behalf of a controller).

There are seven fundamental principles that you need to be aware of:

- Lawfulness, fairness and transparency;
- Purpose limitation;
- Data minimisation;
- Accuracy;
- Storage limitation;
- Integrity and confidentiality;
- Accountability.

There are also eight individual rights that you need to be aware of:

- Right to be informed;
- Right of access;
- Right to rectification;
- Right to erasure;

- Right to restrict processing;
- Right to data portability;
- Right to object;
- Rights concerning automated decision-making and profiling.

The penalty for non-compliance is a fine up to the greater of £17.5 million or 4% of global annual turnover.

The Privacy and Electronic Communications Regulations 2003 (as amended) provide additional regulation for e-communications, including electronic marketing and website cookies. Consent from recipients for these to be received must be provided expressly (i.e., opt-in check box).

Additional legislation comes in the form of the Freedom of Information Act 2000. This gives the public the right to request information held by public authorities. If a written request is submitted, the public body must provide a written response within 20 working days. This can give the information (potentially with a fee payable) or refuse to provide the information, including an explanation.

When an instruction is completed, you must ensure that your file is organised and stored for a sufficient period. Under the Limitation Act 1980, claims can be brought up to six years from the date of the breach (i.e., when the service was provided or the loss was suffered) and extended to 12 years if the contract was made under deed. Claims can also be brought up to 12 years from the date of knowledge of the damage, subject to a 15-year long stop date. You should check your firm's policy on data retention, which may vary for different types of data or documents.

You may be asked to keep client, instruction or project data confidential via a confidentiality agreement (CA) or non-disclosure agreement (NDA). These will set out the specific restrictions and data the agreement applies to.

Finally, Intellectual Property (IP) is something that you need to be aware of. This covers copyright, trademarks, design rights and moral rights. Copyright is the most common form of IP you will likely come across in your role as a surveyor. This means that the author of the original work has exclusive rights to control the distribution of their work. This right can be licensed, transferred or assigned to another party.

For example, the surveyor owns the copyright to a valuation report. However, the client is licensed to use the report in connection with the purpose of the report. Copyright in this scenario would be breached if the report were to be published elsewhere by the client without the consent of the surveyor (owner of the copyright). Another example is where an architect owns the copyright over their design documents. This must be considered when a developer purchases a site and intends to amend or use the original designs for marketing purposes.

How Can You Demonstrate Level 1?

Your assessors will start at the highest level you have declared, which will be level 1. This means that you only need to state your knowledge (level 1) and do not need to include any examples for this competency in your summary of experience.

Typical questions will start with, 'Tell me about your knowledge of . . .', 'What do you know about . . . ?' or 'Explain . . .'. You can substitute any data management principles outlined in this chapter to write yourself a set of sample questions.

Reference List

RICS, 2017. *Professional Statement Conflicts of Interest*. [Online] Available at: www.rics.org/content/dam/ricsglobal/documents/standards/january_2022_conflicts_of_interest_global_1st_edition.pdf [Accessed 14 July 2023].

Chapter 10

Diversity, Inclusion and Teamworking

Diversity, Inclusion and Teamworking is a mandatory RICS APC competency required to level 1. This means that you need to write up your summary of experience with a statement of knowledge at level 1 only.

The RICS pathway guide sets out this competency's relevant knowledge (level 1).

All surveyors will work in internal and external teams during their careers, so clearly understanding what makes a team effective and perform well is essential. Various theories of teamworking are helpful to comprehend, including Tuckman's Stages of Group Development.

You need to understand how good communication improves team cohesion and collaboration. You can read more about communication in this book's Communication and Negotiation chapter.

You must also be aware of, respect and promote diversity and inclusion, in line with Rule 4 of the RICS Rules of Conduct (RICS, 2021). This states that 'Members and Firms must treat others with respect and encourage diversity and inclusion'.

This chapter will explain what you should know concerning Diversity, Inclusion and Teamworking, covering the main requirements of level 1.

Types of Teams

Surveyors will work in both internal and external teams during their careers. Internal teams could be formed at several levels, including company-wide, department-specific or project teams. External teams will include external personnel, such as clients, planners, solicitors, lawyers, architects, structural engineers and other surveyors.

DOI: 10.1201/9781003437116-10

Teams can be permanent or temporary. The latter will be formed around a project, instruction or steering group rather than being a permanent team within a company.

Team Formation and Roles

Tuckman's model is one of the fundamental models of teamwork. It explains how teams develop and function using a five-stage process:

- Forming – where team members are still working as individuals rather than as a team. In this initial stage, team members get to know each other and how the team will function;
- Storming – as the team develops, it will encounter various teething problems and conflicts. These include challenging the team's objectives, leadership, workload, progress and ways of working;
- Norming – as the team begins to resolve differences and overcome conflict, it will function more smoothly and operate more collaboratively towards common objectives;
- Performing is where the team functions effectively with structured processes and commitment from each team member;
- Mourning – this will be the end of the team, perhaps due to a project or instruction being concluded.

Teams may progress and regress throughout the five stages of this model, so it is not a one-way street. This could happen when a new team member joins or a challenge is encountered on a project.

It is a good exercise to think about a team that you have worked in or are working in. What stage is the team in, or what stages did the team work through during a project or instruction? How could the progress of the team have been improved?

Teamwork Skills

Think about some of the best and worst teams you have worked in. What do you think made them effective?

Here are some of the things that you may have come up with:

- Having a common purpose and clear objectives;
- Having sufficient resources, including a sufficient budget and personnel (with a range of skills);

- Team members having mutual respect and an understanding of each other's strengths and weaknesses;
- Knowledge and expertise being shared openly;
- Team members being able to speak out openly;
- Having a combination of different personality styles among the team members.

If you have experienced a team that did not function well, why do you think this was the case? What barriers or challenges did the team face?

Some of the most common barriers to effective teamwork include:

- Inadequate resources;
- Misunderstanding of objectives;
- Poor selection of members;
- The wrong mix of skills and personalities;
- Poor leadership;
- Wrong size (typically, teams of between five and eight people tend to work most effectively);
- Inadequate training.

It is important for each team member to understand their role within the team. Belbin's theory of team roles is a good way to do this. Belbin defined a team role as 'a tendency to behave, contribute and interrelate with others in a particular way' (Belbin, 2015). Belbin split nine overall team roles into three categories (Figure 10.1).

Each role will have its own strengths and weaknesses, and an effective team is likely to have a wide variety of roles rather than lots of team members who occupy the same role. Which role do you think you take on within a team? How might having an understanding of your team role help you to work more effectively in future teams?

Another way to define your personality and how you might behave in a team, is to take a Myers-Briggs Type Indicator test. You might have taken one of these as part of the application process for a job or as part of an assessment centre. Why not see what one of these tests online says about you?

A helpful model is Maslow's Hierarchy of Needs, which seeks to explain the motivations behind human behaviour. Basic needs at the bottom of the pyramid need to be met before those higher up can be attained (see Figure 10.2):

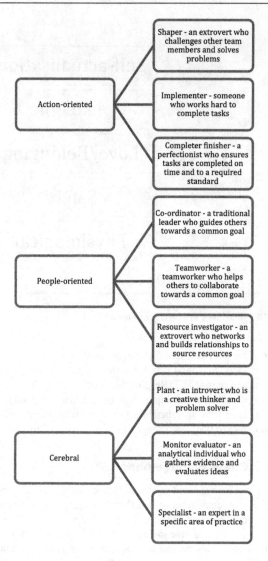

Figure 10.1 Belbin's Team Roles

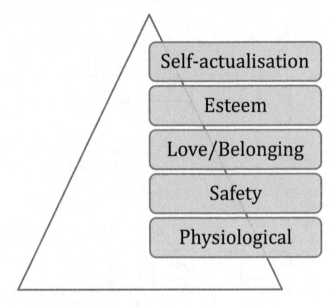

Figure 10.2 Maslow's Hierarchy of Needs

A RACI matrix is a tool that can be used to manage the responsibility of individuals and teams for specific tasks within a project. It uses the acronym RACI (Responsible, Accountable, Consulted and Informed) to define ownership of each task. A matrix is then created, allocating each team member to a task or activity, along with their ownership role of that task, using one of the four RACI roles.

Supply Chain Management

Supply chain management relates to how businesses supply and buy services and products from one another. Managing the supply chain in a positive, collaborative and honest way provides short- and long-term benefits (e.g., cost and time savings) for all parties involved.

In relation to this competency, teamworking is an essential part of supply chain management, particularly when working with an external team of suppliers to deliver an overall project or instruction.

Collaboration from the outset helps to avoid design issues, claims and delays while ensuring that all suppliers have the right information available at the right time.

Strategic alliances can be an important part of supply chain management, where a long-term relationship is formed between companies with clearly defined goals and mutual benefits.

Applying the principles learnt elsewhere in this chapter can help you to improve your supply chain management.

Diversity and Inclusion

RICS is 'committed to raising awareness and promoting diversity and equality for the profession where every individual has an opportunity to thrive and fulfil their potential' (RICS, 2023).

This sits comfortably with Rule 4 of the RICS Rules of Conduct, which states that 'Members and Firms must treat others with respect and encourage diversity and inclusion'.

The Equality Act 2010 is the primary UK legislation relating to discrimination based on nine protected characteristics:

- Age;
- Gender reassignment;
- Being in a marriage or civil partnership;
- Being pregnant or on maternity leave;
- Disability;
- Race;
- Religion or belief;
- Sex;
- Sexual orientation.

Discrimination comes in a variety of forms, including direct (being treated less favourably than others), indirect (having rules that apply to all but which put those with a protected characteristic at an unfair disadvantage), harassment (being treated in a way that violates your dignity or is offensive) and victimisation (being unfairly treated because you have raised a complaint about discrimination or harassment).

Discrimination can also come in the form of unconscious bias, whereby an individual's views and life experiences influence how they treat or interact with others. For example, it could mean treating someone favourably because they are similar to you or treating some-one less favourably because they are different to you. Unconscious bias also includes stereotyping, where a generalisation (which is often not true) is applied to an individual or group of people. Being aware of your own unconscious bias is the first step to tackling it. Firms can challenge unconscious bias in recruitment by listing roles in multiple places, blind sifting (omitting certain details on application forms) and ensuring interview panels include a diverse range of individuals.

Historically, there has been a lack of diversity in the property and construction industries. CIOB (2023) reported the construction work-force in 2019 comprised circa 12% female workers, 5–7% BAME workers and 2.7% LGBTQ+ workers. This underrepresentation of the wider make-up of society will have an impact on the outputs from the industry and is something that continues to be tackled by a variety of different means.

Neurodiversity is also an important consideration, relating to the differences in how individuals process and interpret information. Examples of neurodiversity include ADHD, autism, dyspraxia, dys-lexia and dyspraxia. All forms of neurodiversity have a profound impact on how we interact with the built environment, how we do our jobs and how we relate to others. Being aware of neurodiversity and incorporating it into how we design spaces (including considering PAS 6463:2022 Design for the Mind), communicate with others and work in teams is essential to promote diverse and inclusive workplaces.

Inclusive communication is a key part of creating an inclusive workplace culture. This means communicating in a way that means that the message is understood by as many people as possible, includ-ing those who are neurodiverse. Examples of inclusive communica-tion include using a wide range of communication methods, using visuals alongside text, being flexible in how services are provided and using appropriate language (e.g., avoiding gendered language or stereotyping).

Most firms will have a published diversity and inclusion policy. For example, you can read CBRE's policy on their website (CBRE, 2023). Read your firm's policy and see how it protects you and others from discrimination and promotes fair opportunities in the workplace.

There has been substantial press in recent years about the gender pay gap. The ONS reported that the overall UK gender pay gap in April 2022 was 8.3% (Office for National Statistics, 2022).

However, the problem appears more prolific in the property and construction industries, as Property Week reported an average hourly gender pay gap of 31.88% in 2022 (Property Week, 2022). Suggestions to reduce the gender pay gap include flexible working, enhanced shared parental pay, training and better retention and recruitment policies.

Companies can strive to achieve various diversity and inclusion standards or certifications, such as ISO 30415 (Human Resource Management – Diversity and Inclusion), Economic Dividends for Gender Equality (EDGE) and the RICS Inclusive Employer Mark.

RICS is also a signatory, alongside the CIOB, ICE, LI, RIBA and RTPI, to the Memorandum of Understanding: Creating a More Diverse, Equitable and Inclusive Built Environment Sector. This aims to create a more diverse, equitable and inclusive property and construction sector.

Surveyors should also be aware of the Modern Slavery Act 2015. Modern slavery is a violation of human rights, including slavery, servitude, human trafficking and forced labour. In the construction industry, this could include unsafe working conditions, forced or unpaid work and inadequate accommodation provision. Modern slavery can be tackled by having appropriate policies on anti-slavery, recruitment, procurement and raising concerns (whistleblowing). Section 54 of the Act requires large organisations, meeting specific requirements to publish an annual statement examining modern slavery.

In conclusion, a diverse and inclusive workplace will have substantial benefits for individuals, teams and the wider business. This includes having a positive impact on business performance, staff retention, reputation, innovation and engagement. However, it is certainly not just about lip service or policy – it is about genuinely, honestly and tirelessly promoting an inclusive and diverse culture that welcomes all, irrespective of who or what we are.

How Can You Demonstrate Level 1?

Your assessors will start at the highest level you have declared, which will be level 1. This means that you only need to state your knowledge (level 1) and do not need to include any examples for this competency in your summary of experience.

Typical questions will start with, 'Tell me about your knowledge of . . . ?', 'What do you know about . . . ?' or 'Explain . . .'.

Reference List

Belbin, 2015. *Team Roles in a Nutshell*. [Online] Available at: www.belbin. com/media/1336/belbin-for-students.pdf [Accessed 27 July 2023].

CBRE, 2023. *Diversity, Equity & Inclusion*. [Online] Available at: www.cbre. co.uk/careers/diversity-equity-inclusion [Accessed 27 July 2023].

CIOB, 2023. *Our Charter on Diversity & Inclusion*. [Online] Available at: www.ciob.org/specialreport/charter/diversityandinclusion   [Accessed   27 July 2023].

Office for National Statistics, 2022. *Gender Pay Gap in the UK: 2022*. [Online] Available at: www.ons.gov.uk/employmentandlabourmarket/peopleinwork/ earningsandworkinghours/bulletins/genderpaygapintheuk/2022   [Accessed 20–23 July 2022].

Property Week, 2022. *It's Time to Close the Gender Pay Gap*. [Online] Available at: www.propertyweek.com/feedback/its-time-to-close-the-gender-pay-gap/5119980.article [Accessed 27 July 2023].

RICS, 2021. *Rules of Conduct*. [Online] Available at: www.rics.org/content/ dam/ricsglobal/documents/standards/2021_roc_en.pdf [Accessed 27 July 2023].

RICS, 2023. *Diversity & Inclusion*. [Online] Available at: https://ww3.rics.org/ uk/en/annual-review/diversity-and-inclusion.html [Accessed 27 July 2023].

Chapter 11

Inclusive Environments

Inclusive Environments is a mandatory RICS APC competency required to level 1. This means that you need to write up your summary of experience with a statement of knowledge at level 1 only.

The RICS pathway guide sets out this competency's relevant knowledge (level 1). This competency is often confused with the Diversity, Inclusion and Teamworking competency. However, they are very different in terms of relevant knowledge.

Candidates must understand the principles and processes used to deliver inclusive environments, recognising the 'diversity of user needs and the requirement to put people (of all ages and abilities) at the heart of the process' (RICS, 2022).

This chapter will explain what you should know concerning Inclusive Environments, covering the main requirements of level 1.

Inclusive Design

RICS (2022) defines an inclusive environment as recognising and accommodating 'differences in the way people use the built and natural environment. It facilitates dignified, equal and intuitive use by everyone. It does not physically or socially separate, discriminate or isolate. It readily accommodates and welcomes diverse user needs'.

The Construction Industry Council (CIC) (Construction Industry Council, 2017) publishes a guide called 'Essential Principles for Built Environment Professionals'. This sets out six principles for creating inclusive environments:

DOI: 10.1201/9781003437116-11

- Contribute to building an inclusive society now and in the future;
- Apply professional and responsible judgement and take a leadership role;
- Apply and integrate the principles of inclusive design from the outset of a project;
- Do more than just comply with legislation and codes;
- Seek multiple views to solve accessibility and inclusivity challenges;
- Acquire the skills, knowledge, understanding and confidence to make inclusion the norm, not the exception.

Creating inclusive environments goes beyond just meeting minimum requirements. It is an ethical and moral requirement required of RICS Members under the Rules of Conduct. These require Members to treat others respectfully (Rule 4) and act in the public interest (Rule 5).

Property and construction professionals may be involved in providing inclusive environments in the planning, design, management and operation stages of a project or the lifecycle of a building. This could be on various scales, from a specific building to a neighbourhood, regional or national scale.

During the planning stage, inclusive design should be built into development proposals from the outset. The National Planning Policy Framework requires that planning policies and decisions should create places that are safe, inclusive, accessible and that promote health and well-being. Many local authorities have incorporated relevant policies into their Local Plans or Supplementary Planning Guidance, e.g., 'Accessible London: Achieving an Inclusive Environment'.

Proposals and designs should not segregate or separate individuals or treat users with particular needs differently. This relates to more than just physical access, such as ramps or lifts – it requires the design to remove barriers and enable all users to enjoy the space equally. Examples include providing prayer and washing facilities or ensuring that buildings are easily navigated with clear signage.

Fundamental design principles include:

- Focusing on people, including consultation with stakeholders and building sustainable communities;
- Acknowledging diversity and difference;
- Offering choice rather than one design solution;
- Providing flexibility of use;

- Providing a positive experience when using spaces, including neurodiversity through signage, lighting, visual contrast and trained staff.

Legislation

Although we have a moral and ethical duty to exceed the minimal legal and regulatory standards relating to inclusive environments, it is still essential to understand what these are.

The Equality Act 2010 is the key primary legislation relating to inclusive environments. We have already covered discrimination and the protected characteristics in the Diversity, Inclusion and Teamworking chapter of this book. Under this Act, a duty is placed on service providers and bodies exercising a public function to make reasonable adjustments where someone is disadvantaged because of a disability. This aims to provide equal access for disabled users to goods, services and premises.

This also applies to controllers (i.e., landlord or management company) of let residential premises and common parts. There is also a duty on agents and landlords not to discriminate based on any protected characteristic when selling or letting properties.

The duty to make reasonable adjustments is anticipatory in relation to the protected characteristic of disability (but not any others). The word 'anticipatory' means that service providers should consider what reasonable adjustments may be required rather than waiting for a request.

Reasonable adjustments can include changing how things are done (e.g., a new rule or policy), providing auxiliary aids (e.g., British Sign Language interpreter or a portable induction loop) or amending physical features (e.g., widening doorways, ramps, lifts, automatic doors or more signage). The word 'reasonable' relates to the adjustments' practicality, effectiveness, cost, extent and disruption.

If a service provider fails to make reasonable adjustments, then the service user can take legal action (i.e., a court claim) under the Act. In the case of *Royal Bank of Scotland Group Plc v Allen* [2009] EWCA Civ 1213, a disabled customer requested that his local bank install a platform lift, as he could not access the premises using the existing stairs. The bank refused and suggested that he use telephone or Internet banking. A claim ensued, where the Court of Appeal held that the bank's suggestion was unreasonable and that the customer should be entitled to use the bank in person, as this provision was available to the general public. It was held that an alternative provision was only acceptable where physical access could not be provided.

The Equality Act 2010 also created the Public Sector Equality Duty (PSED). This extends the duty of public bodies to consider how their policies and decision-making affect individuals under the broader protected characteristics of the Act. This includes eliminating unlawful discrimination, advancing equal opportunities and encouraging good relations with the general public.

Approved Document M (Volumes 1 and 2) of the Building Regulations relates to the minimum requirements for the access to and use of buildings. This includes requirements for door widths, lifts and WCs, for example.

Under Regulation 9, there is a ten-year exemption to make reasonable adjustments to physical features under the Equality Act 2010, where the design meets the minimum standards of Approved Document M. Meeting the requirements of Approved Document M, however, only provides minimum requirements and does not guarantee that buildings will be accessible to all. Approved Documents K (protection from falling, collision and impact) and B (fire safety) are also essential to understand in relation to this competency.

Other legislation includes the Regulatory Reform (Fire Safety) Order 2005, Fire Safety Act 2021 and Building Safety Act 2021. Collectively, these require the Responsible Person (i.e., the person in control of the premises) to prepare a fire risk assessment, including the needs of any disabled users. Measures that could be installed include smoke alarms with a vibrating pad or flashing light, smoke alarms with a strobe light or an emergency call system. A Personal Emergency Evacuation Plan (PEEP) should also be prepared for users who may have difficulty exiting a building in an emergency. You may wish to read the Government guidance, 'Fire Safety Risk Assessment: Means of Escape for Disabled People' (2007).

On a more general scale, there are two British Standards relating to inclusive environments, BS 8300:2009+A1:2010 and BS 926:2013, and an ISO, ISO 9999:2016 (assistive products for persons with disabilities). There are various campaigns, such as Changing Places, promoting larger accessible toilets, and certifications, such as Fitwel (a healthy building certification system).

How Can You Demonstrate Level 1?

Your assessors will start at the highest level you have declared, which will be level 1. This means that you only need to state your knowledge (level 1) and do not need to include any examples for this competency in your summary of experience.

Typical questions will start with, 'Tell me about your knowledge of . . .', 'What do you know about . . . ?' or 'Explain . . .'.

Reference List

Construction Industry Council, 2017. *Essential Principles Guide*. [Online] Available at: www.cic.org.uk/shop/essential-principles-guide-for-built-environment-professionals-creating-an-accessible-and-inclusive-environment [Accessed 13 July 2023].

RICS, 2022. *Sector Pathways: Competencies Guide*. [Online] Available at: www.rics.org/content/dam/ricsglobal/documents/join-rics/pathway_guides_requirements_and_competencies.pdf [Accessed 13 July 2023].

Royal Bank of Scotland Group Plc v Allen [2009] EWCA Civ 1213 (2009).

Chapter 12

Sustainability

Sustainability is a mandatory RICS APC competency required to level 1. This means that you need to write up your summary of experience with a statement of knowledge at level 1 only.

The RICS pathway guide sets out this competency's relevant knowledge (level 1). This focuses on balancing environmental, social and economic objectives at various scales. You need to have a robust understanding of this topic, particularly as there is wide-ranging legislation and regulations of which you must be aware. Given the wide breadth of RICS AssocRICS and APC pathways, you need to know about sustainability issues relevant to your practice area.

This chapter will explain what you should know concerning Sustainability, covering the main requirements of level 1.

Sustainability

In 1987, the Brundtland Report defined sustainable development as meeting 'the needs of the present without compromising the ability of future generations to meet their own needs' (United Nations, 1987).

The triple bottom line principles built upon this, defining three objectives of sustainability: environmental, social and economic.

The concept of Corporate Social Responsibility (CSR) takes the triple bottom line and aims to make businesses accountable for their actions. It further proposes that the triple bottom line principles underpin business value and performance.

Environmental and Social Governance (ESG) is a recent concept that builds on CSR. This provides specific criteria that businesses can use to measure their sustainability impact.

DOI: 10.1201/9781003437116-12

In the RICS Professional Standard Sustainability and ESG in Commercial Property Valuation and Strategy (3rd Edition) (RICS, 2023), ESG is defined as,

> The criteria that together establish the framework for assessing the impact of the sustainability and ethical practices of a company on its financial performance and operations. ESG comprises three pillars: environmental, social and governance, which collectively contribute to effective performance, with positive benefits for the wider markets, society and world as a whole.

Many large businesses report on ESG and incorporate it into their core business objectives and aims. This provides a commitment to sustainability to investors and customers and provides more comprehensive benefits to society.

'E' (environmental) relates to climate change, carbon emissions, water pollution, air pollution and deforestation. 'S' (social) refers to matters such as customer satisfaction, data management, inclusion, diversity and community relations. 'G' (governance) relates to the composition of boards, executive remuneration, political contributions and recruitment best practice.

Each of these has different measurable metrics, such as a reduction in carbon emissions (environmental), health and well-being (social) and policies (governance). There are also disclosure regulations such as the Sustainable Finance Disclosure Regulation (SFDR) and the Taxonomy Regulations.

The World Economic Forum has proposed centring the various metrics around four P principles: principles of governance, planet, people and prosperity.

CSR is, therefore, about accountability, and ESG is about measurability.

The circular economy (or cradle-to-cradle) is a concept that focuses on sustainable consumption, minimising waste, reusing or recycling materials and extending the lifecycle of materials and products.

Legislation

The 17 UN Sustainable Development Goals (SDGs), part of the 2030 Agenda for Sustainable Development, aim to address economic, social, environmental and governance challenges.

They are:

1. No poverty;
2. Zero hunger;
3. Good health and well-being;
4. Quality education;
5. Gender equality;
6. Clean water and sanitation;
7. Affordable and clean energy;
8. Decent work and economic growth;
9. Industry innovation and new infrastructure;
10. Reduced inequalities;
11. Sustainable cities and communities;
12. Responsible consumption and production;
13. Climate action;
14. Life below water;
15. Life on land;
16. Peace, justice and strong institutions;
17. Partnership for the goals.

The RICS has committed to working towards the SDGs in its Value the Planet campaign. This includes professional guidance, research, CPD, better data, sustainable finance, thought leadership and strategic partnerships. The RICS is also committed to the UN Global Compact and ten fundamental principles centred around human rights, labour, environment and anti-corruption.

The RICS publishes an annual Sustainability Report. We recommend that you read the latest version via the RICS website as it will likely be a hot topic at your interview.

Global initiatives and campaigns include the UN Race to Zero and the World Green Building Council's Net Zero Carbon Buildings Commitment.

The Climate Change Act 2008 (as amended) requires the UK to reduce greenhouse gas emissions by at least 100% of 1990 levels by 2050. This is otherwise known as the net zero target, set out expressly in the Energy White Paper (2020) and supported by legislation such as the Environment Act 2021 and strategies such as the Heat & Buildings Strategy. This builds on the Paris Agreement (2015) and the original Kyoto Protocol (1997 and 2013–2020).

The Energy Savings Opportunity Scheme (ESOS) is a mandatory energy assessment scheme for large UK companies that meet specific requirements. ESOS requires energy audits for buildings, industrial processes and transport to identify energy-saving measures. The third ESOS deadline falls in December 2023.

The former Carbon Reduction Commitment (CRC) was replaced by the Streamlined Energy and Carbon Reporting (SECR) regime in 2019. It requires large companies meeting specific requirements to report annually on their carbon emissions, energy use and energy efficiency measures.

Carbon emissions are split into three categories: Scope 1 (direct emissions of a company), 2 (indirect emissions related to the company) and 3 (indirect emissions related to the broader supply chain of a company).

Companies may also have to pay the Climate Change Levy, introduced to replace lost revenue from the CRC. Qualifying businesses must pay the CCL on electricity, gas and solid fuel costs.

Certifications and Assessments

There are various certifications and assessments relating to sustainability and energy efficiency.

Energy Performance Certificates (EPCs) were introduced in April 2008 and are required when a building is built, sold or let. An EPC provides information about a building's energy rating, on a scale of A (most efficient) to G (least efficient), energy use, typical energy costs and recommendations to reduce energy consumption.

EPCs are valid for ten years and must be provided to prospective tenants or purchasers at no cost. An EPC must be provided within a maximum of 28 days of marketing commencing (seven days plus 21 days grace period). The penalty for failing to provide an EPC for a domestic property is £200, and for a non-domestic property is 12.5% of the rateable value, subject to a collar of £500 and cap of £5,000 (or £750 if this formula cannot be applied).

Using the Government website, checking if a building has an EPC is easy.

EPCs must be displayed in non-domestic premises over 500 sq m that the general public visits.

EPCs are not required where the building is:

- A place of worship;
- For temporary use under two years;
- A standalone building with total useful floor space under 50 sq m;

- An industrial site, workshop or a non-residential agricultural building with low energy use;
- A building that is due to be demolished;
- Holiday accommodation rented out for less than four months per year or let under a licence to occupy;
- A listed building if the work would unacceptably alter the character of the building. This is a complex area, and specialist advice should be sought if in doubt;
- A residential building used for less than four months of the year.

Display Energy Certificates provide information on the energy use and carbon emissions of public buildings over 250 sq m, which are visited by the general public. They also use a rating of A to G and have an accompanying advisory report. DECs are valid for ten years if the total useful floor area is 250 sq m to 1,000 sq m, or one year if over 1,000 sq m (the advisory report, however, is valid for seven years).

The Minimum Energy Efficiency Standard (MEES) requires all leases of rented domestic and non-domestic property to have a minimum EPC rating of E.

There are several instances where MEES compliance is not required, e.g., where a building is exempt from having an EPC (see earlier), a building does not have an EPC (or it is over ten years old), the tenancy is under six months without security of tenure or the tenancy is over 99 years.

Furthermore, there are specific exemptions that can be registered where MEES compliance is not required:

- Golden Rule – the EPC rating cannot be improved to E or above, despite all improvements being made with a payback of seven years or less (or for non-domestic property, if the works cost more than £3,500);
- Devaluation – the required improvements would reduce market value by over 5%;
- Third party consent is refused by a tenant, superior landlord or planning authority, or conditionality cannot be reasonably met by the landlord.

Exemptions last for five years and cannot be transferred upon sale.

Surveyors also need to be aware of the Building Regulations, specifically Approved Document L (L1A, L1B, L2A and L2B) relating to the conservation of fuel and power for new buildings and refurbishments. Changes to Part L were made in June 2022 in pursuance of the Government's net zero target and the Future Homes and Future Buildings Standards (in 2025). The changes require new homes to produce 31% less carbon emissions than under previous requirements. The Future Homes Standards increase this to require new domestic dwellings from 2025 to produce 75–80% less carbon emissions than under existing standards. The Future Buildings Standard relates to non-domestic buildings being zero-carbon ready.

Other relevant Approved Documents include F (ventilation), O (overheating) and S (infrastructure for charging electric vehicles).

Various voluntary accreditations or certifications are available to recognise a building's sustainability or energy efficiency credentials.

The Building Research Establishment Environmental Assessment Method (BREEAM) assesses the environmental impact of a building across its lifecycle, including the design, specification, construction and operational stages. It can be used for both new developments and refurbishments. BREEAM uses performance benchmarks for categories including energy, land use & ecology, water, health & well-being, pollution, transport, materials, waste and management. Assessments are carried out at the design (interim certificate) and post-construction (final certificate) stages, with a rating awarded from Unclassified (under 30%), Pass (30–44%), Good (45%–54%), Very Good (55–69%), Excellent (70–85%) and Outstanding (85% and over). Less than 1% of buildings in the UK achieve a BREEAM Outstanding rating. The Government's Construction Strategy requires an Excellent rating on all public projects, whilst specific BREEAM ratings may be required as part of local plans or planning conditions for particular developments.

Other certifications include the US Leadership in Energy & Environmental Design (LEED) and Energy Star schemes, Australian Green Star, global WELL certification, NABERS UK (office buildings), Passivhaus (very low energy buildings), Global Real Estate Sustainability Benchmark (GRESB), PAS 2035 (energy efficient retrofitting of domestic dwellings), Build Off-site Property Assurance Scheme (BOPAS) (modern methods of construction) and SKArating (for fit-out projects).

Organisations may also seek to implement ISO 14001 (environmental management system) and 50001 (energy management). These are externally audited and work on the principles of plan, do, check and act. Organisations can also seek B Corp Certification, meaning a business meets high social and environmental performance standards.

Green lease clauses or memoranda of understanding can incorporate sustainability principles or energy efficiency objectives into commercial leases. These can help to agree on shared goals and determine who pays for and receives the benefits. The Better Buildings Partnership (BBP) Green Lease Toolkit is an excellent starting point to read more about this topic.

Sustainable Design, Resource Efficiency and Renewable Energy

On a project or development-specific level, the National Planning Policy Framework (NPPF) provides guidance on achieving sustainable development. In particular, there is a presumption in favour of sustainable development.

The Infrastructure Act 2015 was legislated to introduce new rights relating to fracking and energy exploitation to meet UK carbon budgets.

From a planning perspective, other key issues to be aware of are biodiversity net gain (BNG), Sustainable Drainage Systems (SuDS) and Environmental Impact Assessments (EIA).

BNG is an approach to land management and development that improves the natural environment. NPPF states that development should promote BNG, with the Environment Act 2021 proposing a minimum 10% gain, amongst other requirements.

Designing SuDS requires stormwater to be managed locally and naturally through attenuation, infiltration and slow conveyance. SuDS are encouraged by both the NPPF and various local planning policies.

Buildings incorporating green design or material features could include:

- Recycled materials;
- Green roofs;
- Renewable energy sources, e.g. ground and air source heat pumps, photovoltaic (PV) panels, solar thermal panels (for hot water systems) and wind turbines;

- Water-saving devices;
- Greywater recycling;
- Locally and sustainably sourced materials (material resource efficiency);
- On-site waste management;
- Designed to take advantage of natural light;
- Designed to optimise solar gain;
- High levels of thermal insulation (although noting controversy over the inappropriate design and installation of spray foam and cavity wall insulation in residential dwellings);
- Access by public transport;
- Bike storage and showers;
- Modern methods of construction or off-site manufacture.

The Building Research Establishment (BRE) produce a Green Guide to Specification, which provides guidance on the environmental impacts of elemental specifications.

There are also various Government grants and schemes to promote renewable energy and the installation of energy-saving measures or green design features. This includes the:

- Help to Heat grant scheme, including the Boiler Upgrade Scheme, Home Upgrade Grant phase 2, Sustainable Warmth Competition, Social Housing Decarbonation Fund and the Energy Company Obligation (ECO);
- The Green Deal scheme, which provides loans to householders to install energy-saving measures, such as insulation, double glazing and draught-proofing;
- Non-Domestic Renewable Heat Incentive (RHI), which subsidises the cost of various renewable heating systems over 20 years;
- Smart Export Guarantee (SEG), which pays small-scale renewable energy generators (e.g., solar PV and wind up to 5MW) for surplus energy sold back to the grid at an agreed tariff;
- Changes to the use of rebated diesel and biofuels, primarily affecting the agricultural sector.

Lenders have also introduced green finance for assets to support sustainability or energy efficiency objectives, such as renewable energy, electric vehicles and energy-efficient retrofits.

RICS Guidance

RICS publishes a wide variety of guidance on sustainability; what you need to know about it depends on your practice area. Examples include:

- Development of a Built Environment Carbon Database alongside other built environment organisations such as RIBA and CIBSE;
- Guidance Note Lifecycle Costing (1st Edition);
- International Building Operation Standard (IBOS), which centres around compliance, economics, functionality, sustainability and performance in strategic decision-making, measurement and management of buildings;
- International Cost Management Standard (ICMS) 3, relating to carbon lifecycle reporting for construction projects;
- Professional Standard Sustainability and ESG in Commercial Property Valuation and Strategic Advice (3rd Edition);
- Professional Statement Whole Life Carbon Assessment for the Built Environment (1st Edition);
- Responsible Business Framework for RICS Regulated Firms and Members.

Surveyors in the construction sector need to understand the difference between Whole Life Cost (WLC) and Life Cycle Cost (LCC). WLC is an economic assessment of all costs and benefits relating to a building over a specified period. LCC, however, refers to costs associated only with construction and operation. WLC could include additional costs such as the cost of the land, building income and other non-construction costs.

Valuers also need to consider sustainability and ESG in property valuation, in line with the aforementioned Professional Standard. This includes understanding whether the evidence for a specific asset type in a particular market supports a green premium (or brown discount).

How Can You Demonstrate Level 1?

Your assessors will start at the highest level you have declared, which will be level 1. This means that you only need to state your knowledge (level 1) and do not need to include any examples for this competency in your summary of experience.

Typical questions will start with, 'Tell me about your knowledge of . . .', 'What do you know about . . . ?' or 'Explain . . .'.

Reference List

RICS, 2023. *Professional Standard Sustainability and ESG in Commercial Property Valuation and Strategic Advice*. [Online] Available at: www.rics.org/content/dam/ricsglobal/documents/standards/Sustainability%20and%20ESG_3rd%20edition_standard%20-%20May%202023.pdf [Accessed 27 July 2023].

United Nations, 1987. *Our Common Future*. [Online] Available at: https://sustainabledevelopment.un.org/content/documents/5987our-common-future.pdf [Accessed 27 July 2023].

Chapter 13

Senior Professional Assessment Competencies

This chapter looks at the three senior professional assessment (SPA) competencies; Leadership, Managing People and Managing Resources. These three competencies must be taken to level 2 for SPA candidates only. This is in addition to the rest of the mandatory competencies in the preceding chapters of this book, which SPA candidates must also satisfy.

The RICS SPA Candidate Guide identifies the relevant knowledge (level 1) and tasks (level 2) for these competencies. SPA candidates do not have to write a summary of experience. However, they must incorporate these competencies into their three case studies, notably the first SPA Case Study. Their final assessment interview will also explore their experience and knowledge of these competencies.

By the time most candidates reach the competence and experience required to enrol in the SPA, they will already have gained advanced knowledge of these competencies, including various leadership and motivation models and theories. This could be through personal development at work or undertaking an MBA or other business qualification. There are excellent reference books already published which provide the depth of knowledge required of a senior professional. This is somewhat outside the scope of this book, which is designed as a revision tool covering the basics. As such, we will cover key topics for each competency but highly recommend referring to key reference texts, some of which we have suggested.

Before proceeding, it is vital to understand the difference between management and leadership. Management relates to the planning, coordinating and controlling tasks carried out by a group of people (team) to achieve set objectives. Leadership, by contrast, links to the ability to influence, empower, inspire and motivate team members to

DOI: 10.1201/9781003437116-13

contribute towards broader goals. Senior professionals are likely to do both and require various skills to be effective managers and leaders.

Leadership

It is recommended to read a good reference book on leadership in business or management. An example is *Leadership in Practice: Theory and Cases in Leadership Character* (Seijts & MacMillan, 2018).

We have already covered the following elsewhere in this book:

- Motivation theory, teamwork and team performance in Diversity, Inclusion and Teamworking;
- Organisation design and types of business in Business Planning and Accounting Principles and Procedures;
- Communication in Communication and Negotiation.

What makes a good leader? We all have different opinions on this, although these are some of the key traits that are common to influential leaders:

- Being motivated and driven and using this to motivate others;
- Believing in the vision of their organisation;
- Being able to communicate with others well;
- Having empathy;
- Being resilient;
- Having a robust moral compass and acting honestly, ethically and with integrity (one of the foundations of the RICS Rules of Conduct);
- Continually learning and being experienced in their field;
- Being self-aware;
- Actively listening;
- Being emotionally intelligent;
- Having confidence in themselves and their decisions;
- Being positive;
- Being realistic;
- Being creative;
- Being strategic and logical in decision-making;
- Being selfless and identifying the talents and skills of others to become leaders;
- Delegating and avoiding micro-management.

Of course, this is not an exhaustive list. Why not list the characteristics or behaviours you demonstrate as a leader?

You could also think of who you admire as a business leader. Why are they so effective in their leadership role?

There are various styles of leadership that you may adopt depending on the situation and your individual preferences.

These include:

- Transformational – focusing on change to drive forward an organisation or a team. This will involve motivating others, creating positive energy and developing team relationships;
- Delegative (laissez-faire) – focusing on letting a competent and skilled team do what they do best. This requires a high level of trust and requires just enough direction to avoid disagreements arising within the group;
- Authoritative (visionary) – focusing on making a plan and motivating others to follow this. This requires the leader to guide, mentor and motivate individual team members, avoiding micromanagement which can create a hostile team environment;
- Transactional – focusing on creating a structured environment with 'carrots' and 'sticks' to lead the team towards common goals;
- Participative (democratic or facilitative) – focusing on listening to team members and involving them in critical decisions. Gaining buy-in for key decisions helps to build accountability and encourages team members to collaborate;
- Servant – focusing on the needs of individual team members and providing what is needed for the team to thrive;
- Autocratic – this is only effective in specific circumstances where a leader needs to make and control a critical business decision without the input of others. This can lead to a lack of buy-in and motivation from the team for obvious reasons.

Which leadership styles have you used? When and why did you use them? Have you used a variety of styles during your career? Which worked better than others, and why?

A communication strategy is how a business communicates with its audience. This should be planned, written down and reviewed regularly.

Key considerations include:

- Purpose of the communication;
- Timing and delivery;
- Communication methods or channels;
- Audience;
- Responsibility of key personnel.

Managing People

It is recommended that you read a good book on managing people and HR in business. An example is *A Short Guide to People Management for HR and Line Managers* (Panagiotakopoulos, 2016). In particular, guidance on HR techniques and performance management can be found in this book.

SPA candidates need to have a robust understanding of key HR legislation. The Chartered Institute of Personnel and Development (CIPD) provides up-to-date summaries on their website.

SPA candidates will be responsible for managing people. This could include employees (with a direct employment contract), contractors and consultants (the latter two of whom will be indirect labour, not employed directly by the firm).

Managers will often develop close working relationships with their teams. However, forming clear boundaries and an effective working relationship is essential. The lines can become blurred when friendships form, and this is an area where people management issues can arise.

It is important to set boundaries and expectations at the outset, including the scope of the employee's role, how issues can be raised and how performance will be managed. Managing people and teams requires regular contact and communication, such as team meetings, one-to-one meetings, informal workplace discussions and physical or online noticeboards.

Staff appraisals will likely be annual, with interim performance management meetings as required. These may include specific goal setting, feedback and staff development. Outcomes should be recorded and then reflected upon at the next meeting.

Giving feedback is a vital part of people management. Ideally, feedback should be specific and help team members improve their performance, develop their skills and boost their confidence.

An excellent way to frame feedback is to give positive feedback, followed by developmental feedback (what could be done differently or better next time) and then positive feedback again to finish. This effectively sandwiches the learning point and can help reduce the perception of the feedback being critical or negative. However, managers will quickly get to know their team members and how they best receive feedback, including the format, delivery, timing and phrasing.

Praise and regular appreciation of team members' work and contributions are essential to creating a high-performing team. Managers are also responsible for developing their team through coaching, personal growth and further training.

Having difficult conversations is also part of the role of being a people manager. These are never easy and something you will learn to become more comfortable with. Examples are issues between team members, poor performance, delivering bad news or dealing with personal problems. Avoiding the problem entirely often makes it far worse, so it is better to tackle difficult conversations when the time is right and before they escalate too far.

It is crucial to prepare for difficult conversations, including:

- Planning an appropriate time and private place;
- Researching the matter and having any evidence to hand in advance;
- Seeking guidance from others, if required;
- Involving a neutral third party in the meeting, if you feel that this may help to facilitate a positive outcome;
- Working through the conversation and what you want from it, including the outcome and how you will approach the matter.

During difficult conversations, the following tips may help:

- Active listening;
- Avoiding becoming overly emotional;
- Taking a break if either party needs it;
- Being considerate and kind to the other person;
- Asking questions to explore the matter;
- Agreeing on a clear outcome, including actions or further support.

Following up the conversation in writing and ensuring that further support or actions are provided or taken can help cement the outcome.

If you doubt your ability to handle difficult conversations or manage difficult team situations, then never hesitate to seek further training or support from your firm. People and emotions are some of the most complex parts of the business to manage, especially when matters become personal.

Knowing your team and ensuring a healthy work-life balance is another part of good people management. Burning out at work and undue workplace stress are both unacceptable; we need to recognise and deal with these if we see them happening. Personal matters often impact the workplace; whilst you may not know the detail of what an employee is going through, supporting them in the right way and taking the pressure off when difficulties are encountered is critical.

The best managers can delegate work effectively rather than micromanaging their team. This will include having an effective system to manage and allocate work and allowing communication between team members and the manager. Peer reviewing and stringent signing-off measures will help maintain high-quality assurance and client care levels. Where employees do not have the right expertise or skills, they should be confident to speak out, and the manager should be there to provide the right resource or support.

Tasks can be broken down into non-urgent, important and critical tasks, with the right human resource allocated accordingly. Delegating tasks to the right people at the right time will leave the manager available to handle their core roles and responsibilities. This often involves less technical work and more people and resource management.

Managers and HR departments in larger firms will be responsible for staff recruitment and induction. All employees should have a written employment contract, and all parties should understand the key terms and implications of the employee's role.

Recruitment requires consideration of staffing needs now and in the future, considering the current team and how this may change over time. New staff may be needed for various reasons, such as new projects, increased workload or specific expertise required. Recruitment can be internal and external, with any job advertisements identifying the role and desired personal qualities, skills and experience. Indirect and direct discrimination should be avoided at all times when recruiting.

Inductions require thought and planning to integrate new employees into the broader team and business. The best inductions are gradual, friendly and structured – ensuring employees feel supported and receive the necessary training. A 'buddy' or mentoring system for new employees can be a great way to onboard them into the business, alongside regular meetings to address any teething problems or challenges that arise. ACAS provide valuable templates that can be used throughout the recruitment and induction journey for new employees on their website.

Managers will also need to be able to deal with various HR requests, including sick leave, annual leave, long-term absence, flexible working requests (particularly after the rise of home working during the pandemic), redundancies, dismissals, disciplinaries and family leave.

When an employee resigns, an exit interview should be held. This will help the business to understand why the employee is leaving and to seek feedback on what can be improved in the future.

Where a grievance is raised, disciplinary action may be required, and a manager will need to investigate and resolve the matter fairly and robustly. Support from the business' HR department may be necessary in this instance.

We recommend reviewing the chapters of this book on Communication and Negotiation and Diversity, Inclusion and Teamworking to learn more. The competency description at level 1 also mentions organisational design, communication strategies and high-performing teams. These have already been covered at length, so they will not be covered again here.

Finally, a core skill for people managers to develop is self-reflection. This helps us to understand what we do well, what we do not do so well and how we can improve our people skills and management approach. Seeking 360 feedback (both up and down from employees to managers) can be helpful, as long as lines of communication are honest, open and transparent. Managers also need the opportunity for their own appraisals, training and personal development, which should not be neglected.

Managing Resources

Businesses cannot operate without resources, materials, equipment and labour. Managers of resources need to use these effectively and efficiently to achieve corporate objectives as set out in the business plan. We will not look at labour in this section, as it has already been covered in the management of people earlier in this chapter.

Material resources typically require the lowest management input. This is because work is primarily office-based and comprises basic materials such as IT equipment, paper and writing equipment.

Equipment resources, such as expensive IT or high-tech specialist surveying equipment, are a little more complex.

Construction companies rely far more on material inputs. However, this would be within the scope of a project manager or a quantity surveyor rather than a senior professional managing resources as part of this competency.

Either way, the key to good resource management is having the right materials and equipment available to the right people, at the right time, at an acceptable cost and with minimal wastage. This involves forward planning and forecasting, considering existing and potential work and staff numbers. Managers may delegate material resource management to an office manager or other employee.

The stages of material control are:

- Identification and specification of requirements;
- Materials ordered, taking into account any lead times, delays or stock shortages;
- Materials checked upon delivery;
- Materials stored securely and appropriately;
- Usage controlled and materials re-ordered when required, to start the process again.

Materials are a basic cost of doing business, so using them efficiently will help to improve overall profitability. Certain materials can also be acquired on bases other than purchasing, such as leasing or hiring. This may affect the business' taxation position and liquidity for higher-value acquisitions. A cost-benefit analysis may be required to advise on the most efficient way to acquire resources and assess when repair or replacement is needed.

We recommend reading the chapters of this book on Accounting Principles and Procedures to learn more about accounting techniques.

Surveyors will also need to manage the costs of production, which, when added together, result in the final price paid by the client. These include:

- Profit;
- Prime costs – direct expenses, material, labour and equipment costs;

- Production overheads – indirect workplace costs;
- Marketing overheads;
- Administration overheads.

This will factor into budgeting, which SPA candidates will be responsible for setting and managing. A budget is a financial target for a set period, typically one year. Budgets can be produced on various scales, including company-wide, departmental, equipment, IT and project-specific.

The process of budgeting can be broken down into the following steps:

- Setting the budget period;
- Forecasting income for the budget period;
- Estimating material, labour (direct, supervision and ancillary resources) and equipment requirements;
- Establishing an operating budget by adding the total expenditure to the profit on the forecasted income.

Budgeting can be challenging, particularly in the current market, with high inflation and variable resource costs. For example, in 2023, energy and construction material prices and interest rates increased dramatically. Reviewing any PESTLE or SWOT analysis undertaken as part of the business planning process may help to set budgets set as accurately as possible.

Budgets must be controlled during the financial year (or an alternative period). This includes planning, publishing, measuring, comparing, reporting and correcting.

How Can You Demonstrate Levels 1 and 2?

During your interview, your assessors will ask you about your three case studies, of which the first SPA Case Study will focus specifically on these three SPA competencies. You will likely be asked about wider examples (at level 2) of demonstrating these competencies during your career.

Typical questions will start with, 'How have you . . . ?', 'Tell me about an example when you . . .' or 'How did you lead your team effectively/manage a financial budget/manage people when . . . ?'.

You may also be asked some level 1 questions, focusing on the knowledge behind your practical level 2 examples. Be prepared to discuss the theory behind your actions and how you applied this to produce the best results regarding leadership, managing people and managing resources.

Reference List

Panagiotakopoulos, A., 2016. *A Short Guide to People Management for HR and Line Managers*. 1st ed. London: Routledge.

Seijts, G. & MacMillan, K., 2018. *Leadership in Practice Theory and Cases in Leadership Character*. 1st ed. London: Routledge.

Chapter 14

Submission Advice

Overview of the APC Assessment

All APC candidates, irrespective of their chosen route, must undergo assessment via a written submission and an online interview. However, the requirements of both elements differ significantly for the senior professional, specialist and academic assessments, which will be considered separately in this chapter.

The written submission element is also slightly different for preliminary review candidates. This is because preliminary review introduces a two-stage written submission process, which will be discussed separately from the final assessment submission. The requirement to undergo preliminary review is set out clearly in the previous chapter.

The online interview is based on the candidate's written submission and lasts one hour. The final decision as to whether a candidate becomes MRICS is solely based on performance in this interview. There is no separate assessment of the written submission, apart from the requirement for preliminary review candidates to pass this element of their written submission process and for final assessment candidates to meet the minimum RICS requirements, e.g., word count and case study date validity.

Final Assessment Submission (Excluding Senior Professional, Specialist and Academic Assessments)

All APC candidates, whether undergoing preliminary review or not, will need to submit a written final assessment submission. This applies to candidates on the structured training and straight to assessment

DOI: 10.1201/9781003437116-14

routes and candidates who have successfully passed preliminary review. Therefore, the final assessment submission is the same for all these routes. However, the final assessment submission is very different in structure and content for the senior professional, specialist or academic assessments.

The written submission will be the product of a substantial preparation period and not something that can be drafted in a matter of weeks. Rushing the written submission will likely lead to a poor-quality document, which will not support the candidate through their online interview. Candidates must ensure they present their submission to the highest written standards, including spelling, grammar and proofreading. Therefore, the final submitted documents must be 'client ready' and written formally and professionally.

The APC final assessment submission includes the following elements:

- Diary (structured training candidates only);
- Summary of Experience;
- Case Study;
- CPD Record;
- Professionalism Module and Test.

Throughout their written work, candidates should be acutely aware of plagiarism. Under no circumstances should a candidate seek to include work or text written by another candidate or copied from a published source unless it is correctly referenced. Any red flags of plagiarism will be identified by RICS using the Turnitin plagiarism detection system, leading to further investigation and potential disciplinary action, including removal from the APC assessment process.

What Does the Structured Training Diary Include, and Who Needs to Complete It?

The diary is only required for candidates on the structured training route. Candidates undergoing preliminary review or proceeding straight to assessment do not need to keep a diary.

Structured training candidates should start to record their diary via the RICS Assessment Platform as soon as they enrol on their APC journey. This is because the number of diary days logged counts towards satisfying the minimum 12 or 24 months structured training requirements. Candidates will also use the diary when compiling their summary of experience, choosing their case study topic and discussing progress during Counsellor meetings.

Candidates must only log time spent on their technical competencies in their diary. This means that they should not record any activities or time spent on their mandatory competencies unless these are also selected as technical competencies.

Mandatory competencies will generally be included within a candidate's daily work rather than being their specific focus at any one time. Examples include communicating during an instruction or working within a project team.

Therefore, time spent on the mandatory competencies will not contribute towards a candidate's minimum structured training requirements. However, they must still be recorded in the summary of experience, which is discussed later in this chapter.

Sufficient detail should be included in a candidate's diary, allowing appropriate examples to be identified to meet the requirements of the chosen competencies at levels 2 and 3. The diary does not form part of the candidate's final written submission, although RICS may request it to be submitted separately before a candidate's online interview. Reasons for this include where RICS identifies anything concerning or contentious within a candidate's submission, e.g., the case study topic does not reflect the candidate's declared competencies or the standard of work in their summary of experience.

Therefore, candidates should ensure that their diary is sufficiently relevant and concise rather than overly detailed and lengthy. It should be written professionally and in a high standard of English.

The following details need to be recorded on the RICS Assessment Platform for each separate diary entry:

- Competency and level relevant to the diary entry;
- Days and start date logged in a minimum of half-day blocks;
- Title of the activity;
- Diary entry, including a brief description of the activity and relevant competencies. This should be as specific as possible to aid the candidate's memory when they draft their written submission or for discussion with their Counsellor.

Candidates can combine their activities or experience within their diary entries into more substantial blocks of time. This may be sensible given the wide range of activities a candidate will likely undertake within a day or week. This might mean a candidate blocks measurement or inspection activities from various projects or instructions into one diary entry. A candidate may also block together time spent on one larger project into one entry to avoid repetition. How candidates split their entries will be directed by their competency choices, as some are much broader topic areas than others. Valuation, for example, is typically wider in scope than Measurement.

What Is the Summary of Experience?

The summary of experience focuses on the candidate's competencies. This includes the mandatory competencies relevant to all candidates and the technical competencies specific to a candidate's chosen pathway. The RICS provides a case study template to download on the RICS Assessment Platform, which should be followed rather than the candidate using their format or structure.

The technical competencies are split into core technical competencies, i.e., a requirement for the specific pathway, and optional technical competencies, i.e., where the candidate can select competencies which best suit their role, knowledge and experience. The chosen competencies will either have a stated achievement level, i.e., Ethics, Rules of Conduct and Professionalism to level 3, or candidates may have a choice of the level attained, e.g. optional technical competencies.

The word counts for the summary of experience are absolute: 1,500 for the mandatory competencies and 4,000 for the technical (core and optional) competencies. There is no leeway to exceed these; a candidate's submission can be rejected outright by RICS for exceeding the word count.

The word count for each section (i.e., mandatory and technical competencies) can be allocated as the candidate wishes between each competency. Generally, assigning a higher word count to the level 3 statements is advisable to ensure sufficient experience and detail can be included.

The summary of experience requires candidates to write a summary of their knowledge at level 1 and experience at levels 2 and 3 for each level of each chosen competency. Unlike AssocRICS, this includes Ethics, Rules of Conduct and Professionalism. Candidates should ensure that they refer directly to the pathway guide and competency descriptors when preparing their summary of experience, as this is what they will be assessed against by their final interview panel.

If a competency is selected to level 3, then candidates must write a statement for each of levels 1, 2 and 3. For level 2, a statement must be provided for levels 1 and 2. At level 1, only a level 1 statement is required. At levels 2 and 3, it is advisable to use subheadings or other appropriate formatting to highlight examples, aiding the candidate and assessment panel in the final assessment interview by signposting specific examples.

At level 1, candidates must explain what they know or have learnt about a competency. This may link to CPD activities or academic learning, e.g., university modules. Given the limited word count, a bullet point list of knowledge could be used to reflect the competency descriptor. However, a carbon copy is not allowable and would constitute plagiarism.

Candidates should ensure that the knowledge they declare is relevant to their experience, as the assessment panel can question anything included in the written submission. Candidates should ensure in level 1 that they demonstrate current knowledge of legislation, RICS guidance, and any relevant hot topics.

At level 2, candidates should include two to three practical examples of their work-based experience. This will discuss how they acted or carried out relevant tasks. The range of appropriate activities or tasks for a specific competency should be related to the pathway guide and competency descriptor, although these are not exhaustive lists.

Candidates should seek to be as specific and refined as possible in their examples. This could be by stating a specific instruction, property or client to help focus their final interview questioning. Candidates should avoid using vague or broad examples as this can make it harder to demonstrate the required level in the final assessment interview. If a project or instruction is very broad, then candidates could focus on a specific aspect of their involvement or advice that meets the relevant competency requirements. For example, measurement of a property the candidate subsequently values or inspection of a site which a candidate later advises on the planning and development potential.

At level 3, candidates should again include two or three practical examples of their work-based experience. These examples should relate to where the candidate has given reasoned advice to a client. Likewise, candidates should use specific examples demonstrating their integral role in a project or instruction.

Candidates should ensure they write examples using the first person, using I, me and my. This ensures that the assessment panel know that the work indicates it is that of the candidate rather than that of a colleague, friend or superior. This also gives the assessment panel the perception that the candidate has provided professional advice and handled instructions competently from start to finish, with minimal supervision.

What Is the Case Study?

The case study provides an in-depth discussion of a specific project or projects with which a candidate has been involved and provided reasoned advice. Candidates should ideally focus only on one project, which helps to keep the detail refined and the project easy to understand by the assessment panel. The choice of project or instruction may include work outside of the candidate's geographical assessment region. However, in the final assessment interview, candidates should be aware of the legislation and guidance relating to both the geographical regions of the case study project and the assessment location.

The choice of project should reflect the candidate's chosen pathway and competencies. However, the case study does not need to include aspects of every competency, which is unlikely to be realistically achievable. The competencies demonstrated should consist of level 3 aspects, clearly showing reasoned advice being given to a client.

The candidate must have been involved in the case study project or instruction within the last two years from the submission date. If a project is out of date, this is a point for referral, and a candidate will not be permitted to proceed to the online interview stage. Some projects, e.g., a construction project or site disposal, may extend for more than two years. However, the candidate's primary involvement must have occurred during the two-year validity period. Including a clear timeline to demonstrate the candidate's involvement in their case study project is helpful. This can also help to demonstrate a candidate's visual communication skills.

The candidate should have played a key role in the case study project. For some instructions, this will include running, managing and leading the project from start to finish. For larger, more complex projects, a candidate may not have been running the project independently but will have played a key role in giving reasoned advice to the client.

Some candidates may have been involved in only one aspect of a project, have been involved after the commencement of a project or have finished their involvement before the project was completed. This is acceptable, provided the candidate can demonstrate their reasoned advice in the stated level 3 competencies. If a project has not been completed when a candidate writes their case study, they may wish to discuss the prognosis within the case study and provide an update in their final interview presentation.

The case study choice does not need to be the most complex, high value or large project or instruction the candidate has been involved with. Given the limited word count, using a highly complex case study that the assessment panel does not understand clearly or fully can be challenging and counterproductive. This can lead to misunderstandings and confused questioning as the candidate's role and advice have not been made clear in the context of the limited word count and written case study structure. The best case studies are often simple but succeed in setting out a clear and logical story of the candidate's involvement and reasoned advice.

Within their project or instruction choice, candidates should identify and discuss two or three key issues involving challenges they encountered and needed to overcome during the instruction. This will demonstrate the candidate's problem-solving and analytical skills, leading to the provision of reasoned advice and recommendations to the client. If a candidate cannot identify at least two key issues, the project or instruction is unlikely to be a suitable case study topic.

Example case study, key issues for a secured lending valuation instruction could include:

- Key issue 1 – Assessment of Market Rent;
- Key issue 2 – Assessment of Market Value, including choice of yield;
- Key issue 3 – Advice on secured lending to the client.

Example case study key issues for a level 3 Building Survey could include:

- Key issue 1 – Defect 1 – wet rot;
- Key issue 2 – Defect 2 – condensation;
- Key issue 3 – Remedial advice to the client.

The word count for the case study is 3,000 words, which is again absolute with no leeway. This includes everything between the end of the contents page and the start of the appendices. Candidates can be referred before being invited to interview by RICS for exceeding the word count. As such, it should be treated as a strict client requirement. Candidates should avoid using the appendices to include additional detail which could not be fitted within the main body of the case study. This does not demonstrate adherence to a client's strict requirements and will not be viewed favourably by the assessment panel.

Rather than using footnotes, candidates should include core references to legislation or RICS guidance within the main body of the case study. This prevents the case study from becoming overcomplicated and untidy.

The RICS provides a case study template to download on the RICS Assessment Platform, which should be followed rather than the candidate using their format or structure. The following sections should be included:

1. Introduction – summary of the project, candidate's role, responsibilities, key stakeholders and timeline.
2. My Approach – this should discuss two to three key issues relating to the project and the challenges the candidate overcame. Some candidates may only use one key issue, but this should be avoided as the level of detail and advice will likely be limited. Trying to include more than three key issues is likely to lead to insufficient detail and analysis, given the relatively limited overall word count. The candidate should explain the issue, the options considered, and the solution, advice and recommendations given. This requires analysis of the advantages and disadvantages of each option or the specific considerations in giving reasoned advice to the client.
3. My Achievements – the candidate should discuss their key achievements, potentially using subheadings for each competency demonstrated. This provides helpful structure and relates the actions and advice to the candidate's competency choices. The candidate should also discuss the outcome of the instruction and their overall achievements in terms of the level 3 reasoned advice they gave to their client.
4. Conclusion – the candidate should reflect on and critically analyse their involvement in their case study project, including lessons learnt and how they will improve their performance in the future. This section requires as much thought as the preceding ones and demonstrates a candidate's commitment to professional development. It is also advisable to discuss the candidate's professionalism

and demonstration of ethics within the case study project, as this is a crucial aspect of being a Chartered Surveyor.

5. Appendices – the candidate should include an initial Appendix A to list the mandatory and technical competencies demonstrated within the case study. The other appendices should consist of only relevant illustrations, photographs or plans, which should be referenced within the main body of the case study. Excessive or irrelevant appendices should not be included.

Confidentiality is a crucial ethical issue for candidates to consider. Candidates must have the express consent of their employer and client to disclose sensitive or confidential details within their submission. Suppose that this is not provided by one or more parties. In that case, specific details should be redacted, e.g., use Project X instead of the address or secured lending client instead of the lender's identity. Anything litigious should also be redacted to ensure that identifiable details, such as the client's name, address or sensitive financial information, are not disclosed. Candidates can be referred for not dealing with confidentiality appropriately, as it relates to a candidate's ethics and professionalism. RICS also confirms that anything in the candidate's submission will not be disclosed further than the assessment panel.

What Does the CPD Record Include?

APC candidates must record at least 48 hours of CPD for every 12-month period. Candidates requiring 24 months of structured training must record 24 months of CPD, with the minimum requirement recorded for each 12-month period. Candidates who need to record 12 months of structured training must record 12 months of CPD, with the minimum requirement recorded for this period.

All other candidates need to record their last 12 months of CPD activities. For preliminary review candidates, 12 months of CPD is required before both the preliminary review submission and the final assessment submission. The latter should be updated to reflect the relevant 12-month period.

Candidates do not need to submit CPD records for a period in excess of this, i.e., there is no requirement to submit CPD for a period longer than the required 12 or 24 months. Submitting an extended CPD record could lead to additional questioning that a candidate could otherwise avoid. Where this relates to historic CPD, the detail and outcomes may be difficult to remember or less relevant to the candidate's current role and experience.

A candidate would be better served by including relevant CPD for the required period and in sufficient depth and detail. In particular, candidates should never forget that their assessment panel can question anything they write in their final assessment submission, including their CPD record.

RICS calculates the CPD requirement on a rolling basis from your submission date. For example, suppose a candidate submitted in February 2023. In that case, their CPD must be undertaken in the 12 or 24 months, if appropriate, before this, i.e., from February 2021 or 2022 to February 2023. The CPD requirements do not relate to the CPD recording requirements for qualified Members, which are lower. This is a common source of confusion for APC candidates.

All CPD logged by APC candidates should relate clearly to the chosen technical and mandatory competencies relevant to the candidate's scope of work, role and responsibilities. Ideally, a candidate will undertake various types of CPD to maximise their learning opportunities. Candidates should also plan their CPD and ensure that they reflect on what they have learnt afterwards, demonstrating their analytical and evaluative skills.

At least 50% of the minimum 48 hours of CPD must be formal. This requires a structured approach with clear learning objectives. Examples of formal CPD (RICS, 2020) include structured seminars, professional courses (which do not have to be run by RICS), structured self-managed learning and online webinars.

CPD is logged by candidates via the RICS Assessment Platform, including the following key details:

- Description of the activity – for example, 'Webinar – JCT Contracts' or 'Self-managed learning on the Red Book';
- Activity status – future or completed;
- Start date;
- Time – hours and minutes allocated to the CPD activity;
- Activity type – formal or informal and method, e.g., webinar, conference, private study or mentoring;
- Learning outcomes – reflection on what was learnt during the CPD activity. This should be written in the first person and past tense, e.g., 'I learnt about the different procurement methods and how

they can be applied in my role', or 'I learnt about the RICS Professional Standard and the core principles which I can apply in the workplace'.

Candidates must provide sufficient detail in their CPD record to allow the assessment panel to understand how the activity relates to a candidate's competencies, roles and experience. If poorly written or incomplete, it may constitute grounds for referral. The same applies if insufficient hours have been undertaken or activities are informal rather than formal, meaning the minimum requirements are unmet. Candidates should also proofread their CPD record carefully and ensure that the detail provided is accurate, relevant and concise.

What Is the Professionalism Module?

Candidates must also complete the online RICS Professionalism Module and Test (accessed via the RICS Assessment Platform) within 12 months of applying for final assessment. The module is also a helpful revision tool which candidates should consider using to prepare for their final assessment interview. Ethics is the only area for automatic failure at the interview stage, so candidates should appreciate the importance and content of this module.

What Is the Preliminary Review Submission?

Candidates who do not have an RICS-accredited degree, i.e., a non-cognate degree, will need to undergo the additional preliminary review submission stage.

This means that a candidate prepares their written preliminary review submission and submits this for a written review by RICS before submitting their final assessment submission roughly four to six months later, or longer if preferred.

The candidate's final assessment and preliminary review submissions do not differ in structure or format. It is simply a 'pre-check' process carried out by RICS for candidates without an RICS-accredited degree, which provides confidence that minimum standards of writing, structure and quality are being met.

The purpose of preliminary review is to determine if a candidate's submission, in terms of content, quality and professionalism, is of the standard RICS requires. Essentially, this allows the candidate's assessment panel to effectively prepare for and conduct the final assessment interview.

The preliminary review assessment panel will consider the following:

- Whether holistically the candidate's submission is of the required standard;
- Whether basic RICS requirements are met, including word count, professionalism, written English, structure and sufficient examples at levels 2 and 3;
- Whether content requirements are met, such as including relevant detail and evidence relating to a candidate's chosen competencies in their summary of experience and case study. This requires the candidate to reflect on the requirements of the pathway guide and competency descriptions within their written submission.

The results for preliminary review candidates typically take around eight weeks to be issued on the last working day of the month. Substantial feedback will be provided within the feedback report whether or not the outcome is a pass or a referral.

If the submission is acceptable, the candidate may submit their final assessment in a future submission window. Candidates and their Counsellors should closely scrutinise their feedback report to ensure they make relevant improvements in their final submission.

Candidates may also wish to update their experience, examples and CPD activities to reflect new work undertaken following their preliminary review result being received. It is worth remembering that the preliminary review feedback is the subjective view of the specific panel at the time; sometimes, it is impossible to make every single amendment, particularly given the opinions or advice of a candidate's Counsellor or employer.

Candidates must also ensure that their case study remains valid (i.e., within a two-year window) when they come forward for final assessment. For example, this may not be the case if the preliminary review stage was passed more than six months prior. In this case, the candidate may need to write a new case study, considering the general feedback provided at preliminary review. If a candidate does change their case study topic or examples included in their summary of experience, they do not need to submit for preliminary review again.

If the submission is not acceptable, the candidate will receive a feedback report and need to resubmit for preliminary review before submitting for final assessment. Again, candidates and their Counsellors should review the report together and ensure improvements are made before the candidate resubmits for re-assessment at a future preliminary review window.

A successful preliminary review outcome is not a guarantee of success at the candidate's final assessment interview. This is because the decision to become MRICS is solely based on the candidate's final interview.

Conclusion

Candidates should be able to apply what they have learnt in this book's preceding chapters when writing their final submission. The next chapter will explain how this will be taken one step further during the final assessment interview.

Reference List

RICS, 2020. *Annex A: Examples of types Formal and Informal CPD Activities.* [Online] Available at: https://www.rics.org/globalassets/rics-website/media/upholding-professional-standards/regulation/media/cpd-annex-a-160518-mb.pdf [Accessed 16 October 2023].

Chapter 15

Interview Advice

What Is the Final Assessment Interview?

After candidates submit their final assessment via the RICS Assessment Platform, they will be notified by email of their final assessment interview date and a minimum of two week's notice beforehand. All interviews are online, and RICS currently plans no further assessments to take place face-to-face.

If a candidate submits their final assessment documents but subsequently wishes to defer until the next sitting, they can do so by emailing RICS. No charge is payable if the deferral is notified within 14 days of submitting. However, a charge will apply thereafter. Candidates may defer for various reasons, e.g., extenuating circumstances that affect their availability or if they feel they are not ready to be assessed.

The assessment panel will consider the following during the candidate's final assessment interview:

- The candidate's communication skills both through a ten-minute presentation on their case study and responses to the assessors' questioning on their final assessment submission;
- How well the candidate can verbalise and explain their advice and actions discussed within their written submission, demonstrating the stated competency levels;
- Whether the candidate understands the role and responsibilities of a Chartered Surveyor, including advising clients diligently and acting ethically and professionally;
- Whether the candidate can act independently and unsupervised, i.e., that they are a safe pair of hands. This is particularly relevant because after qualifying as MRICS, a candidate could, in theory,

DOI: 10.1201/9781003437116-15

set up in practice as a sole trader. The assessment panel must be confident that any successful candidates in their APC would be competent to do this ethically and professionally.

The assessors will assess the candidate's written submission, although this will be considered holistically within the interview context. Unless there is an evidence deficiency or issue regarding word count or basic requirements of the written submission (e.g., CPD hours), a candidate will not be referred on their written submission alone. Any substantial issues, such as the word count being exceeded, should be picked up by RICS after the candidate submits via the RICS Assessment Platform, and the candidate would then not be permitted to proceed to the final interview stage.

Candidates will be assessed by a panel of two or three Chartered Surveyors, one chairperson, and one or two assessors. All will be trained APC assessors, or chairs, and experienced in providing final assessment interviews. To ensure that the interview is tailored to the candidate's experience, two of the panel members will be from the same pathway as the candidate. Some pathways also have specialist areas, and at least one panel member will be experienced in this area.

Assessors are trained to give every candidate a fair interview process and to treat all candidates with respect and dignity. The interview process should encourage and facilitate all candidates to succeed if they meet the required levels of competence. The online assessment process means that an even more diverse range of assessors is available to participate in the assessment process, which is likely to give candidates a much more positive, fair and transparent interview experience.

Candidates should be made to feel at ease by their panel, who should be positive and provide candidates with an environment in which they can excel. Ways they may do this include:

- Asking open questions, one at a time, to avoid being confusing or overly elaborate;
- Being flexible in their questioning approach and following up on a candidate's answers with further questions if appropriate;
- Supporting a candidate and encouraging them if they are nervous or stressed.

However, panels will not give candidates any indication of how they are doing during the interview or whether they have been passed or referred.

Therefore, the best indication of a correct answer is sometimes the panel moving on to another competency or question. If an assessor is pursuing the same line of questioning or driving at a specific point, this may indicate that the candidate's answers need further thought or clarity. At this point, asking for clarification or returning to the question at the end of the interview may be sensible.

A staff facilitator may also be present during the interview to assist the candidate and assessment panel with any technical issues. An auditor may also be present to observe the assessment panel's performance, not the candidate's. They are simply there to assess whether the panel are undertaking the assessment in accordance with RICS standards. They do not take any part in the panel's decision-making on your performance.

The assessment panel should have no perceived or actual conflicts of interest in assessing the candidate fairly and transparently. This could be either personal, e.g., a candidate and assessor having met at a CPD event or being familiar with each other in a professional context, or prejudicial, e.g., where an assessor may benefit from the candidate's success. A personal conflict may not present an issue if both parties are happy to proceed. However, a prejudicial conflict should be declared and the assessor removed from the panel. If a candidate becomes aware of a conflict of interest on the day, they should make their chairperson aware. A two-person panel proceeding will deal with the matter or the interview being rescheduled with a non-conflicted panel.

Candidates should have made RICS aware of any access arrangements (formerly known as special considerations) or extenuating circumstances during the submission process. RICS may request supporting medical evidence to enable relevant, reasonable adjustments to be made. Suppose a candidate feels the panel may benefit from knowing how specific issues affect them. In that case, they may wish to prepare a written statement to read out to the panel during the initial welcome at the start of the interview.

The interview lasts 60 minutes and is conducted online using Microsoft Teams. Candidates should ensure they test the system beforehand and have a good-quality video camera and microphone. There is no excuse for hardware or systems not working on the day.

Candidates can prepare effectively for the interview by:

- Choosing a quiet interview location, where disturbances are kept to a minimum. This may mean asking others to keep quiet or to leave the location for the duration of the interview. If a candidate's home environment is too noisy, then they may prefer to sit their interview in an office or external meeting room;

- Having good quality Wi-Fi or tethering to 5G or 4G if not available. This should be tested beforehand, and if weak, a candidate should reconsider their choice of location or ask other users to switch their devices off during the interview;
- Having good lighting and a background that is neutral and clear. Candidates could sit near natural light or have a lamp beside or behind them. They should avoid sitting in front of a window which can be blinding to the camera;
- Ensuring their interview environment is comfortable, including a supportive chair, adequate height desk, notepad, pen and a glass of water. Candidates may also choose to use a second screen, such as a tablet or monitor, to use for their presentation notes only;
- Using headphones or a separate microphone to ensure good quality audio;
- Avoiding using a smartphone which can be too small to allow an effective interface during the interview;
- Closing any applications not being used to minimise the risk of technical issues, as well as putting any devices on silent or do not disturb;
- Ensuring their camera is at eye height to keep a natural connection with the assessment panel;
- Wearing smart, professional clothing appropriate for a client meeting or job interview. Ideally, this should be simple and plain to avoid being distracting on camera;
- Ensuring that any devices are connected to a power source rather than relying on battery power which is quickly drained.

A candidate is not permitted to record their interview. Any attempt to do so may lead to disciplinary action and immediate termination of the interview.

The 60-minute time limit is strict and will be managed closely by the chairperson. In the event of technical difficulties, up to ten minutes can be added to the end of the interview. If interruptions total more than ten minutes, the chairperson will terminate the interview, which will be rescheduled.

A candidate's interview may also be extended if any special considerations or extenuating circumstances dictate that additional time should be given to allow the candidate to respond fully to the questions asked. In this case, the candidate will not be asked more questions than they would otherwise be in a standard 60-minute interview. Instead, it is the amount of time they are given to listen, comprehend and respond to questions that may be extended.

Candidates should join their interview video link five to ten minutes beforehand. They will then be permitted access from the virtual lobby when their assessment panel is ready.

The interview will be structured as follows:

- Initial welcome by the chairperson before the 60-minute time limit starts. This will include the candidate being asked to complete a 360 view of their surroundings to ensure that no outside assistance is being provided. The chairperson is free to ask for this to be repeated at any time during the interview. The chairperson will then explain the interview structure and ask if the candidate is fit, well and ready to proceed;
- Ten-minute presentation by the candidate, focusing on their case study;
- Ten minutes of questioning by the assessors on the candidate's case study;
- Thirty minutes of questioning and discussion by the assessors on the candidate's summary of experience, CPD, Rules of Conduct and professional practice;
- Ten minutes of questioning by the chairperson on any outstanding competencies and ethics. This will include the chairperson's closing comments and the opportunity for the candidate to have the last word. This may consist of returning to any questions the candidate was unable to answer earlier on or making any additional comments, which can be noted down on a piece of paper during the interview.

The candidate's case study presentation could focus on one or more of the key issues. However, it should not simply repeat verbatim what a candidate has written. The presentation could be made interesting by delving deeper into one of the key issues, looking at the progress of the project since the case study was written or discussing further the candidate's analysis of the options and advice given to the client. The main aim of the presentation is to demonstrate strong communication and presentation skills to the panel. Therefore, the presentation does not need to be complicated, nor does it necessarily need to introduce anything new to the panel.

A visual aid can support the presentation. This can be screen shared with the panel.

The best visual aids are generally simple and used only to support key points within the presentation. This could be a simple PowerPoint presentation or a series of PDF slides. Avoiding animation or overly complex slides is advised.

Using a flipchart or visual aid physically held up to the camera is not recommended, as this is likely to be hard to read by the assessment panel. Visual aids should be professionally presented, with a clear title and large, easy-to-read text or graphics. Given the reliance on IT technology to screen share, candidates should be prepared not to use a visual aid in the event of technical difficulties.

Candidates can use cue cards or brief notes for their presentations. During the rest of the interview, they cannot use any notes or have a copy submission to hand.

The timing of the case study presentation is key, and candidates will be asked to stop if they exceed this by more than ten to 15 seconds. Equally, being substantially below this will be a negative considera-tion in the context of the assessors' overall decision. Candidates should practise their presentation frequently to ensure they are fluent in giving it and can accurately meet the timing requirement. A stopwatch, clock or timer can be used, providing that it is not overly distracting to the candidate and their assessment panel.

Given that the interview is conducted online, candidates must best use both verbal and non-verbal communication. Candidates should aim to speak clearly and concisely, taking time to understand and listen to questions before answering. They should also be aware of avoid-ing too much movement or gesticulation, which can be distracting on camera. Positive body language will help to portray a confident can-didate, and practising this on camera beforehand with a friend, family member or colleague can be helpful. Candidates should always try to look directly into the camera and maintain good eye contact with their assessment panel.

The interview is based on the candidate's written submission. The assessment panel can question anything included in the candidate's case study, summary of experience and CPD record. Therefore, candi-dates may wish to limit references to complex case law, for example, in their written submission, unless they are confident to discuss these in their interview. Candidates should also be aware that they will be assessed concerning the legislation and guidance relating to the geographic region of their final assessment. If experience is declared from other countries, candidates must know the legislative and regulatory requirements of each jurisdiction.

Candidates may also be asked about current hot topics regarding industry or market issues, provided they are relevant to the candidate's area of practice. This means that having a good level of market awareness is essential, which can be obtained through reading trade press and a good quality newspaper, listening to relevant podcasts and watching CPD videos.

Assessors are trained to begin questioning at the highest level declared, with supporting level 1 knowledge-based questions potentially being asked to explore the justification for the advice or actions of the candidate. Candidates should ensure they are familiar with any examples in levels 2 and 3, as these should form the basis of their answers. Candidates will not be asked questions on competencies they have not selected or at levels beyond those declared, i.e., they will not be expected to give reasoned advice (level 3) if they have declared a competency only to level 2 (acting or doing).

Questions posed by the panel should not be hypothetical, and they should encourage the candidate to answer based on their experience. If possible, all competencies will be questioned by the assessment panel, with ethics questions included within the main body of the interview. The chairperson will also ask additional ethics questions in the final ten minutes. Candidates should be aware that elements of their mandatory competencies may be assessed within their technical competency questioning. A candidate's communication skills will also be evaluated during their case study presentation and response style to the panel's questions.

Assessors are trained to signpost candidates to the competency area being questioned. Candidates should ask for clarification if they are unsure of the question or the area of focus being sought by the assessor. Candidates should also ensure they listen carefully to the questions they are being asked, take a deep breath before answering and then seek to give only the response the assessors require.

The overall assessment decision is holistic; one wrong answer (unless related to ethics) will not constitute a referral. This means that poor performance in one competency may be balanced out by good performance elsewhere. However, an inability to demonstrate several competencies to the required level will likely constitute a referral, particularly if these are the competencies declared by the candidate at level 3. Candidates are not expected to be experts in every area of their professional practice. The assessors are, therefore, seeking to confirm that the candidate has met the minimum required levels of competence declared.

Candidates should also be mindful that acting ethically is at the heart of being a Chartered Surveyor. Ethics, Rules of Conduct and Professionalism is the only competency where a wrong or unethical answer will constitute an automatic referral.

The assessment is not designed to be an exam. It is an assessment of professional competence, and there will be questions that a candidate cannot answer. This requires a toolbox of potential responses to be practised by candidates. For example, some questions may relate to experience or knowledge outside the candidate's core scope of competence or practice. For these questions, the candidate can state this and identify where they would seek specialist advice, input or support from. This shows that the candidate can take responsibility and is a safe pair of hands when dealing with clients.

Candidates may also encounter questions they simply cannot answer due to the pressure of the interview process. In these situations, referring or signposting the assessors to suitable RICS guidance or documents is helpful. When under pressure, candidates should avoid explaining everything they know about a topic or issue to the assessors. This does not show diligence or a considered approach to advising clients. Instead, a candidate would be better placed to return to the question later or use some of the aforementioned tools and tactics.

The assessment panel will decide on the final result immediately after the interview. In the case of a three-person panel, the assessors will make a decision, and if a split decision is reached, the chairperson will make the final decision. The same applies to a two-person panel, where the chair and assessor will make their own decisions, although the chairperson has the final say.

What Does the Senior Professional Submission and Assessment Include?

The senior professional assessment is one of three separate APC assessment routes. This requires candidates to have at least ten years of relevant experience, which is reduced to five years if a postgraduate degree is held. Candidates also need to demonstrate senior professional responsibilities in their role, including leadership, management of people and management of resources.

The RICS defines a senior professional as 'an individual with advanced responsibilities who is recognised for their impact and career progression within the profession' (RICS, 2020a). This requires demonstrating the qualities (or indicators) of leadership, managing people and managing resources.

Therefore, the senior professional assessment is appropriate for candidates in senior management positions or those responsible for managing or leading teams rather than being responsible for day-to-day surveying work. Indicators of this assessment route being appropriate are a candidate's standing in their organisation's structure, having decision-making responsibilities, an international dimension to their role, a high-profile client base and recognition from the wider industry, peers and media.

Senior professionals should also be able to demonstrate at least one of the following behaviours in their work:

- Pursuing opportunities to develop the industry and the profession;
- Advocating best practice;
- Taking responsibility to deliver professionalism;
- Acting with integrity to promote responsible business.

Senior professional candidates need to satisfy each of the three senior professional competencies to level 2 and the mandatory and technical (core and optional) competencies relevant to their chosen pathway.

The senior professional competencies are:

- Leadership;
- Managing people;
- Managing resources.

Senior professionals must undergo an initial vetting stage by submitting an application form to RICS, including their employment history, qualifications, pathway and 400-word senior profile statement.

The 400-word senior profile statement should discuss the candidate's role, activities and senior professional responsibilities, referencing the aforementioned indicators and behaviours. The candidate should also include relevant detail on their activities and impact as a senior professional, with a supporting organogram showing their senior professional position in their organisation.

Successful candidates are eligible to enrol for the senior professional assessment and must submit their final assessment submission within 12 months. If this deadline is missed, the candidate must re-apply for the initial vetting stage by RICS.

The senior professional written final assessment submission includes the following elements:

- Application form, as submitted at the initial vetting stage;
- The CPD record, although the requirements differ from the 'traditional' APC routes. Candidates require 20 hours of CPD for the last 12 months before submitting, including at least 50% comprising formal activities;
- RICS Professionalism Module and test certificate, which needs to be dated in the 12 months before submitting for final assessment;
- Competency and pathway selection;
- Three case studies of 1,000 to 1,500 words each.

The three case studies each focus on a single project or instruction, where the candidate has demonstrated their senior professional role in management, leadership, client relationship management or strategic advice. The candidate may have delegated or overseen the provision of some of the technical input by employees, consultants or contractors.

The three case studies have different requirements:

- Senior professional case study – focusing on a project demonstrating the senior professional competencies;
- Technical case studies 1 and 2 – focusing on different projects in each case study that demonstrate the candidate's experience against at least two core technical competencies, of which at least one must be shown to level 3, i.e., where they have given reasoned advice. These case studies should also demonstrate the three senior professional competencies, relevant mandatory competencies and the candidate's appreciation of ethics and professionalism.

When selecting appropriate case study projects, candidates should bear in mind the following:

- The case study must have been undertaken within the last three years, calculated on a rolling basis from the submission date. Some projects will have extended over a longer period, in which case the

candidate's primary involvement must have been in the last three years;
- The candidate should include at least one case study which has been undertaken in the country they will be assessed in, allowing them to demonstrate relevant knowledge of national legislation and guidance;
- The project could have been provided to an internal or external client;
- Candidates need to have express client and employer's consent to use the chosen case studies or, if this is not possible, to redact any details which make the project identifiable.

Each case study should follow a consistent structure, including the following elements:

- Overview of the project, objectives and key issues;
- Role in the project, including responsibilities, actions and reasoned advice;
- Analysis of key issues, challenges or problems and recommended solutions or remedial action. This will focus on the candidate's approach and delivery of key objectives or outcomes;
- Conclusion looking at the key achievements and broader impacts on the client, employer, candidate's career and future work;
- Statement of specific competencies which have been demonstrated;
- Appendices, such as drawings, photographs or plans. These should not be extensive and should be kept to only those directly relevant to the case study content.

The case studies should be carefully proofread to demonstrate high standards of professionalism.

After submitting the final assessment submission, candidates will be invited to a 60-minute online final assessment interview by RICS. This aims to assess whether the candidate has:

- Applied their level 1 knowledge through professional experience and advice (levels 2 and 3) and recognised the impact of their senior professional role and responsibilities on the relevant sector or market. This requires the candidate to demonstrate clear comprehension of their selected competencies and pathway;
- Acted ethically and in accordance with their duty of care to clients, employers and the wider public;

- Acted as an ambassador for the profession;
- Acted in pursuance of the objectives of their client or employer;
- Demonstrated understanding of up-to-date and geographically relevant legislation and technical theory.

The 60-minute interview is structured as follows:

- Ten minutes – the candidate's presentation, providing a personal introduction and background on their senior professional career history and current role. This should not focus specifically on the case studies, as is required for the 'traditional' APC assessment;
- Fifty minutes – the questioning and discussion of the candidate's submission, senior professional role, responsibilities and broader professional issues, including ethics.

The weighting of the panel's decision will be allocated as 50% towards their senior professional profile, 25% towards the pathway competencies and 25% towards ethics and professionalism.

The interview is not an exam, and the assessment panel will structure it as a professional discussion, focusing on exploring the candidate's senior professional role and background. The questioning will be based on the submitted documents but with broader issues investigated to ensure that the candidate has reached the level required of a Chartered Surveyor.

There will be an appreciation and understanding by the assessment panel that senior professional candidates will often be managing or delegating work to others rather than undertaking day-to-day technical work themselves. This means that the emphasis will be on the candidate's management and leadership skills, alongside a strong focus on ethics and professionalism, particularly when leading or managing other professionals, contractors or consultants.

What Does the Specialist Submission and Assessment Include?

Similar to the senior professional assessment, the specialist assessment requires candidates to have at least ten years of relevant experience, which is reduced to five years if a relevant undergraduate degree or relevant professional qualification and a relevant postgraduate degree (master's or higher) are held.

Specialist candidates need advanced responsibilities in a specialist or niche area of work. The RICS defines a specialist as 'an individual delivering enhanced services who is recognised for their impact and authority within the profession' (RICS, 2020b).

Being a specialist may also include the following indicators:

- Having a decision-making position;
- A track record of specialist consultancy work;
- Speaking at conferences;
- Writing in the trade press;
- Being appointed by a governance/judicial body;
- Being recognised by the wider industry, peers and media;
- Lecturing;
- Providing formal training;
- Being qualified above master's level (e.g., PhD);
- Being involved in dispute resolution for a technical area.

Specialists should also be able to demonstrate one or more of the following behaviours in their work:

- Pursuing opportunities to develop the industry and the profession;
- Advocating best practice;
- Taking responsibility to deliver professionalism;
- Acting with integrity to promote responsible business.

The initial stage of applying for the specialist assessment involves undergoing the vetting stage. This requires completing the specialist application form, including the template CV, chosen APC pathway and 400-word specialist profile statement.

When choosing a relevant pathway, specialist candidates should ensure they can satisfy one or two core technical competencies relating to their specialist area of work. At least one of these must be declared at level 3.

The specialist CV requires the candidate to confirm their academic education, professional qualifications and professional experience (including an overview of the scope and responsibilities of each role). It also requires the candidate to confirm the specialist indicators demonstrated in their work from the following:

- Position in the organisation structure;
- Publications;

- Record of specialist consultancy work;
- Record of speaking at high-level conferences;
- Dispute resolution;
- Recognition;
- Appointment by governance or judicial body;
- Record of lecturing or formal training;
- Qualifications.

The specialist profile statement of 400 words should provide a clear overview of why the candidate is eligible for the specialist assessment. The candidate should discuss their specialist services and activities, together with the behaviours that demonstrate the authority and impact of their specialist role. The statement should include an organogram or description of the specialist's role in their organisation's structure.

RICS will review the vetting form against the following criteria:

- Sufficient experience;
- Relevant qualifications;
- Experience in the chosen pathway;
- Reference made to the required specialist indicators and at least one specialist behaviour;
- Evidence of specialist authority, enhanced services and outcomes/impact.

Successful candidates are eligible to enrol for the specialist assessment and must submit their final assessment submission within 12 months. If this deadline is missed, the candidate must re-apply for the initial vetting stage by RICS.

The specialist-written final assessment submission includes the following elements:

- Application form, as submitted at the initial vetting stage;
- CPD record, although the requirements differ from the 'traditional' APC routes. Candidates require 20 hours' CPD for the last 12 months before submitting, including at least 50% comprising formal activities;
- RICS Professionalism Module and test certificate, which needs to be dated in the 12 months before submitting for final assessment;

- Competency and pathway selection;
- Three case studies.

The three case studies have different requirements, although all need to relate to the candidate's specialist profile:

- Specialist case study – focusing on a project demonstrating specialist experience in one or two of the candidate's chosen technical core competencies, of which at least one must be to level 3 (i.e., giving reasoned advice);
- Technical case studies 1 and 2 – focusing on different projects in each case study that demonstrate the candidate's experience against at least two technical competencies.

Each case study has a word count of 1,000 to 1,500 and should focus on different technical case studies. The case studies should also reference the mandatory competencies, particularly Ethics, Rules of Conduct and Professionalism, where appropriate.

When selecting appropriate case study projects, candidates should bear in mind the following requirements and considerations:

- The case study must have been undertaken within the last three years, calculated on a rolling basis from the submission date. Some projects will have extended over a longer period, in which case the candidate's primary involvement must have been in the last three years;
- The candidate should include at least one case study which has been undertaken in the country in which they will be assessed, allowing them to demonstrate relevant knowledge of national legislation and guidance;
- The project could have been provided to an internal or external client;
- The candidate should have led the project and been involved in setting strategy, decision-making, analysing options, recommending solutions and managing the client relationship;
- Some of the 'day-to-day' technical work may have been delegated to employees, consultants or contractors;
- Candidates need to have the express consent of their client and employer to use the chosen case studies or, if this is not possible, to redact any details which make the project identifiable.

Each case study should follow a consistent structure, including:

- An overview of the project, objectives and key issues;
- The role of the specialist in the project, including responsibilities, actions and reasoned advice;
- An analysis of key issues, challenges or problems faced and the solutions or remedial action recommended. This will focus on the candidate's approach and delivery of key objectives or outcomes;
- A conclusion looking at the key achievements and wider impacts on the client, employer, candidate's career and future work;
- A statement of specific competencies which have been demonstrated;
- Appendices, such as drawings, photographs or plans. These should not be extensive and should be kept to only those directly relevant to the case study content.

The case studies should be carefully proofread to demonstrate high standards of professionalism.

After submitting the final assessment submission, candidates will be invited to a 60-minute, online, final assessment interview by RICS. This aims to assess whether the candidate has:

- Applied their level 1 knowledge through professional experience and advice (levels 2 and 3) and recognised the impact of their work on the relevant sector or market. This requires the candidate to have understood and demonstrated clear comprehension of their selected competencies and pathway;
- Acted ethically and in accordance with their duty of care to clients, employers and the wider public;
- Acted as an ambassador for the profession;
- Acted in pursuance of the objectives of their client or employer;
- Demonstrated understanding of up-to-date and geographically relevant legislation and technical theory.

The 60-minute interview is structured as follows:

- Ten minutes – the candidate's presentation providing a personal introduction and background on their specialist work and case studies;
- Fifty minutes – questioning and discussion of the candidate's submission, specialist area of work and wider professional issues, including ethics.

The weighting of the panel's decision will be 50% towards their specialist profile, 25% towards the pathway competencies and 25% towards ethics and professionalism.

The interview is not an exam, and the assessment panel will structure it as a professional discussion, focusing on exploring the candidate's specialist role and background. The questioning will be based on the submitted documents but with wider issues investigated to ensure that the candidate has reached the level required of a Chartered Surveyor.

What Does the Academic Submission and Assessment Include?

The academic assessment is appropriate for academic professionals, e.g., lecturers or researchers. The assessment requirements reflect the differences in how competence will be demonstrated by academics as opposed to practising surveyors.

Eligible academics must have at least three years of academic experience and a surveying-related degree. The academic experience does not have to be continuous and can be undertaken at various times.

Academic candidates must satisfy the same mandatory competencies as all other APC candidates, irrespective of route or pathway. Academics must also select a pathway aligned to their area of academic practice, e.g., Commercial Real Estate or Quantity Surveying and Construction. This will inform the candidate's technical competency choices, where at least one level 3 core competency from the chosen pathway must be selected alongside other core competencies applicable to all academic candidates. These include Data Management to level 2 and either Research Methodologies and Techniques or Leadership to level 3. The selection between the latter two will depend on whether or not the candidate has leadership and management experience (i.e., Leadership to level 3) or if their role is more focused on data collection and analysis (i.e., Research Methodologies and Techniques or Leadership to level 3).

Academic candidates must initially submit their CVs to RICS using the academic CV template available on the RICS website. This includes the following:

- Personal details;
- Business details;

- Pathway selection;
- Education history;
- Professional qualifications;
- Professional experience for at least the last three years, including employers, roles and an overview of the candidate's scope and responsibilities;
- Academic criteria;
- Academic review statement relating to the four pieces of evidence selected above.

The academic review statement comprises a 3,000-word summary of the candidate's academic experience and how four pieces of relevant evidence relate to their chosen pathway and the wider surveying profession. The four pieces of evidence do not need to be submitted to RICS at this time. These only need to be presented with the candidate's final assessment submission.

Competence needs to be demonstrated within three main areas, with candidates submitting four pieces of evidence in total from the following three lists to support their application:

- Teaching – preparation and delivery of learning material, assessment of undergraduate and postgraduate student work through summative and formative marking and feedback, completion of a postgraduate teaching qualification, fellowship of the Higher Education Academy, mentoring and supervision of research students and course leadership and development;
- Research and Scholarship – publishing built environment or surveying-related research in journals, conference proceedings, consultancy reports, Government research, legal reports and books. Candidates should ensure that explicit references are included for any publications mentioned, which should be recent, relevant and of national significance or quality;
- External Engagement and Academic Activities – including embedding research, employability or professional practice into undergraduate or postgraduate level curricula, industry engagement and knowledge transfer. This could include student liaison groups, guest lectures, CPD events or engaging with an RICS Committee or Board.

As discussed, at least one piece of evidence must evidence level 3 in one of the candidate's core competencies. Furthermore, at least one piece of evidence needs to come from each of the categories that follow, with the fourth piece of evidence coming from any of the categories depending on the candidate's area of academic focus. Candidates must have express consent to include the evidence in their employer and clients' submissions.

The academic review statement should be structured as a professional report, using subheadings reflecting the following recommended sections:

- Introduction – approximately 100 words explaining the candidate's career history, academic profile and role;
- Teaching (Evidence 1) – approximately 700 words explaining how the candidate's teaching experience is relevant to their pathway and the surveying profession;
- Research and Scholarship (Evidence 2) – approximately 700 words explaining how the candidate's research and scholarship experience are relevant to their pathway and the surveying profession;
- External Engagement and Academic Activities (Evidence 3) – approximately 700 words explaining how the candidate's external engagement and academic activities are relevant to their pathway and the surveying profession;
- Additional Item (Evidence 4) – approximately 700 words explaining how the candidate's final piece of evidence is relevant to their pathway and the surveying profession.

RICS will first review the CV and academic review statement to ensure the academic pathway is appropriate for the candidate. If approved, these two documents will be further reviewed by an academic review panel, which will confirm if the candidate can proceed to final assessment based on the quality and content of the documents submitted. The academic review panel will consider the candidate's eligibility based on their academic profile rather than assessing their professional competence, which will be assessed in the final assessment interview.

The academic review panel report and feedback will be issued within 21 days. If the candidate is unsuccessful, then RICS will suggest alternative routes that would better fit the candidate's experience and role.

If a candidate is successful at the second stage review, they can submit for final assessment and interview.

The final assessment submission should include the following:

- CV, as submitted at the initial review stage;
- Academic review statement, as presented at the initial review stage;
- Four pieces of supporting evidence, which have not been submitted at the initial review stage, although they will have been discussed in the candidate's academic review statement. The evidence should relate closely to the candidate's mandatory and technical (core and option) competencies;
- Summary of experience;
- CPD record – the requirements are the same as for the 'traditional' APC routes, i.e., 48 hours for the last 12 months before submitting, unlike for the senior professional and specialist assessments;
- RICS Professionalism Module and Test certificate – which must be completed in the 12 months before submitting for final assessment.

For the summary of experience, candidates should refer to the afore-mentioned detail relating to the 'traditional' APC routes. The requirements are very similar, although focusing on the candidate's academic role and experience for the academic pathway. In particular, candidates should ensure that at level 3, they provide sufficient detail and practical experience relating to their academic practice at higher education level. The word counts are 1,500 for the mandatory competencies and 3,000–4,000 for the technical (core and optional) competencies. This roughly equates to 150–200 words per level and competency.

Candidates who submit for final assessment will subsequently be invited for an online interview by RICS. This lasts for 60 minutes with a panel of two or three assessors, at least one of whom will specialise in the academic route.

The interview will be structured as follows:

- Five minutes – chairperson's opening and introduction;
- Ten minutes – candidate's presentation on their career, role and one of the four pieces of evidence, explicitly highlighting the candidate's academic skills. A visual aid can be used; candidates should refer to the interview advice earlier on in this chapter;
- Fifteen minutes – questioning on the candidate's presentation;

- Fifteen minutes – questioning and discussion on the candidate's wider submission and academic role;
- Ten minutes – questioning and discussion on CPD, Rules of Conduct and professional practice;
- Five minutes – chairperson's close and opportunity for the candidate to make any final comments or requests for clarification.

The panel will specifically be assessing the candidate concerning the following criteria:

- Breadth of relevant academic experience and knowledge and transference of this to students, researchers or external consultants;
- Requirements of the chosen mandatory and technical (core and optional) competencies and wider pathway context;
- Oral and written communication skills through the written submission, presentation and interview responses and discussion;
- Understanding of the role and responsibilities of a Chartered Surveyor;
- Ethical and professional attitude with a clear duty of care provided to clients, employers and wider stakeholders;
- Being a good ambassador for the profession.

How Do Candidates Receive Their Results?

All candidates will receive notification of their assessment result within five working days of their final assessment interview via the RICS Assessment Platform and shortly after via email.

For senior professional, specialist and academic candidates, the result may take longer, generally seven days for senior professionals and specialists and 21 days for academics.

After qualifying, candidates will be added to the RICS Global Members Directory within 24 hours, a public announcement will be placed on the RICS website a few days later, and an Award Pack will be issued by post six to eight weeks later. Candidates must also pay an upgrade fee to reflect their MRICS qualification status. Concessions may apply to certain professionals, and RICS can advise on eligibility, for example, academics working in academia rather than in a surveying role.

Conclusion

After reading this chapter, candidates should feel confident applying what they have learnt throughout this book to their final assessment interview.

Reference List

RICS, 2020a. *Senior Professional Assessment Applicant Guide*. London: RICS.
RICS, 2020b. *Specialist Assessment Applicant Guide*. London: RICS.

Index

Note: Page numbers in *italics* indicate figures, **bold** indicate tables on the corresponding page.

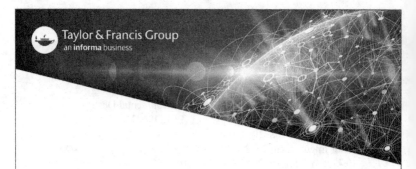

Printed in the United States
by Baker & Taylor Publisher Services